Instructor's Guide for Use with

Laboratory Exercises in
MICROBIOLOGY

Jamie S. Colomé
California Polytechnic State University

Raúl J. Cano
California Polytechnic State University

A. Mark Kubinski
California Polytechnic State University

David V. Grady
California Polytechnic State University

West Publishing Company
St. Paul   New York   Los Angeles   San Francisco

CONTENTS

PART I
MICROSCOPIC TECHNIQUES

EXERCISE 1
USE AND CARE OF THE MICROSCOPE

Exercise 1 introduces the student to the basic skills involved in the proper use and care of the compound microscope. We recommend that prepared slides be used so that students can focus on the mechanics and optics of the compound microscope rather than on techniques for preparing slides. Both parts A and B should be performed in the same period, but part B ought not be performed until the student is well acquainted with the operation of the microscope.

OBJECTIVES:

A.   Develop skills in the use of the compound microscope;

B.   Learn the proper maintenance and care of the microscope;

C.   Develop an appreciation for microbial diversity by viewing the prepared slides.

REQUIREMENTS FOR EACH LABORATORY SECTION

	QUANTITY	COMMENTS
Microscope	1/S*	Exercise should be performed by each individual student.
Prepared slides	6 sets/lab	To include bacteria, algae, protozoa, and fungi. These can be obtained from supply houses (see Appendix) or from slides collected from previous classes.
Lens paper	1 pad/S	To be supplied by student.
Immersion oil	1 bottle/2S	Low fluorescence preferred.

*1/S indicates 1 per student; 1/2S indicates 1 per 2 students.

EXERCISE 2
MEASURING CELLS USING THE MICROSCOPE

Exercise 2 introduces the student to the technique of measuring microorganisms using an ocular micrometer. Although the procedure in this exercise includes instructions for installing the ocular micrometer, we recommend that the micrometer be inserted into the ocular lens by the instructor in order to avoid possible damage to the microscope or introducing dust into the tube. This exercise can be performed in conjunction with exercise 3, having the student measure the same organism (preferably a bacterium) in stained and unstained preparations. The instructor should encourage the student to report the average and the range of width and length of microbial cell measurements.

OBJECTIVES:

A.   Learn the procedure for calibrating the ocular micrometer;

B.   Develop skills in measuring microbial cells;

C.   Illustrate the concept of intraspecies variation;

D.   Illustrate size differences between eukaryotes and prokaryotes.

REQUIREMENTS FOR EACH LABORATORY SECTION

	QUANTITY	COMMENTS
Microscope	1/S	
Ocular micrometer	1/S	
Stage micrometer	1/2S	Lab partners can share one stage micrometer.
Prepared slides	6 sets/lab	To include bacteria, algae, protozoa, and fungi. These can be obtained from supply houses (see Appendix) or from slides collected from previous classes.
Lens paper	1 pad/S	To be supplied by student.
Immersion oil	1 bottle/2S	Low fluorescence preferred.

EXERCISE 3
OBSERVATION OF LIVING MICROORGANISMS

Exercise 3 introduces the student to two techniques of viewing living microorganisms. We recommend that the student examine wet mounts and hanging drop preparations of motile and nonmotile bacteria as well as protozoa and algae (including diatoms). This way the student has the opportunity to practice his skill in microscopy as well as those in the preparation of the slides for microscopy. Both parts A and B should be performed in the same period.

OBJECTIVES:

A. Learn how to prepare wet mounts;

B. Learn how to make hanging drop preparations;

C. View motility and Brownian motion by microorganisms;

D. Practice viewing unstained specimens.

REQUIREMENTS FOR EACH LABORATORY SECTION

	QUANTITY	COMMENTS
Microscope	1/S	
Pond water	4 jars/lab	Should contain a wide variety of microorganisms.
Pepper infusion	4 jars/lab	A 1:2 mixture of black pepper and water in a flask left covered at room temperature for 7 days will make an excellent source of motile bacteria.
Glass slides	1 box/S	To be supplied by student.
Coverslips	1 box/S	To be supplied by student.
Pasteur pipets	1 box/lab	
Depression slide	1/S	To be supplied by student.
Vaseline	1 jar/lab	
Toothpicks	1 box/S	
Lens paper	1 pad/S	To be supplied by student.
Immersion oil	1 bottle/2S	Low fluorescence preferred.

EXERCISE 4
SPECIALIZED LIGHT MICROSCOPES

Exercise 4 introduces the student to specialized microscopic techniques using modifications of the bright field microscope. We recommend that this exercise be performed as a demonstration, perhaps in conjunction with exercise 1. Additionally, viewing the same organism, e.g., an alga like <u>Spirogyra</u>, with the bright field microscope as well as with the dark field, phase contrast, and the fluorescence microscopes provides the student an excellent opportunity for comparing the various types of microscopes. The use of fluorescent dyes is not necessary to demonstrate fluorescence but it might be interesting for the student to note that fluorescence microscopy can be used to diagnose microbial diseases. Any commercially-available kit can be used (same ones used in exercise 71C).

OBJECTIVES:

A.   View microorganisms with the aid of dark field, phase contrast, and fluorescence microscopes;

B.   Compare the uses and limitations of these microscopes.

REQUIREMENTS FOR EACH LABORATORY SECTION

	QUANTITY	COMMENTS
Microscopes:		
A. bright field	1/S	
B. dark field	1/lab	Prepare a demonstration.
C. phase contrast	1/lab	Prepare a demonstration.
D. fluorescence	1/lab	Prepare a demonstration.
Pond water	4 jars/lab	Should contain a wide variety of microorganisms.
Pepper infusion	4 jars/lab	A 1:2 mixture of black pepper and water in a flask left covered at room temperature for 7 days will make an excellent source of motile bacteria.
Cultures:		
A. Sacharomyces	1/4S	On Sabourad Dextrose broth.
Prepared slides:		
A. Borrelia	1/2S	
B. Bacteria	1/2S	Stained with fluorochromes.
Glass slides	1 box/S	To be suppled by student.
Coverslips	1 box/S	To be supplied by student.
Protective glasses	1/S	Must be provided to or purchased by the student.
Lens paper	1 pad/S	To be supplied by student.
Immersion oil	1 bottle/2S	Low fluorescence required.

PART II
CULTIVATION OF
MICROORGANISMS

EXERCISE 5
PREPARATION OF CULTURE MEDIA

Exercise 5 familiarizes the student with the various steps and procedures involved in the preparation of culture media. This exercise also can be used to illustrate the differences between chemically-defined and complex (undefined) culture media. We recommend that students work in pairs to economize on time and supplies and still have each student participate actively in the preparation of the media. Groups larger than four will usually have one or two "observers." It might be advisable to direct the student to exercise 7D for a brief discussion on the proper procedure for pipeting liquids. Although mouth pipeting can be done, it may be better to use pipeting devices so that the students gets in the habit of using such devices for all pipeting done in the laboratory.

OBJECTIVES:

A. Practice the techniques of preparing culture media;

B. Illustrate the differences and similarities of chemically defined and complex media;

C.   Familiarize the student with some of the properties and uses of agar;

D.   Expose the students to some of the shortcomings that occur in the preparation of culture media;

E.   Provide the student with the experience of using the autoclave.

REQUIREMENTS FOR EACH LABORATORY SECTION

	QUANTITY	COMMENTS
Erlenmeyer flask	1/2S	500 ml, with closures
Graduated cylinder	1/2S	250 ml
Pipets (TD)	1/S	1 ml and 10 ml
Balance	4/lab	Should be able to measure 0.0005 g with reasonable accuracy.
Weighing paper	1 box/lab	
Hot plate	1/2S	A ringstand with an asbestos pad may also be used. The student should be reminded to constantly stir the agar mixture to avoid scorching and watch that it does not boil over.
Stirring rod	1/2S	
Autoclave	1/lab	
Heat-proof gloves	1 pair/2S	
Culture tubes	10/S	18x150 borosilicate glass, with heat-proof closures.
Petri dishes	5/2S	Pre-sterilized
Spatulas	10/2S	Wooden, disposable spatulas or popsicle sticks are suitable for dispensing the various reagents. Use one for each reagent.

	QUANTITY	COMMENTS
Chemicals:		
Agar	1 jar/lab	
Glucose	1 jar/lab	
NH_4Cl	1 jar/lab	
K_2HPO_4	1 jar/lab	
$MgSO_4 \cdot 7H_2O$	1 jar/lab	
$FeCl_3 \cdot 6H_2O$	1 jar/lab	
NaCl	1 jar/lab	
Beef extract	1 jar/lab	
Yeast extract	1 jar/lab	
Peptone	1 jar/lab	
Water bath	2/lab	Set at 45° to 55°C

EXERCISE 6
MICROORGANISMS IN
THE LABORATORY ENVIRONMENT

Exercise 6 illustrates to the students that their environment is laden with microoranisms, which can, under appropriate conditions, contaminate their cultures. We suggest that this exercise be performed along with exercise 5 and that the media prepared in that chapter be used in this exercise. Alternatively, exercise 6 can be performed along with chapter 7 in order to emphasize the importance of aseptic techniques in preventing contamination of cultures. The air plates can be saved so that the student can perform various staining procedures (e.g., Gram and endospore stains) outlined in part III. Perform exercise in pairs.

OBJECTIVES:

A. Illusrate that the laboratory environment has many potential sources of microbial contaminants;

B. Emphasize to the student that aseptic techniques are necessary to prevent contamination;

C. Practice observing and describing various microbial colonies.

REQUIREMENTS FOR EACH LABORATORY SECTION

	QUANTITY	COMMENTS
Culture media Nutrient agar	3/2S	18x150 tubes containing 20ml of presterilized nutrient agar are ideal. The plates prepared in exercise 5 can also be used.
Petri dishes	3/2S	
Sterile water	1/2S	1-2 ml in screw-capped tubes
Cotton swabs	1/2S	
Boiling water bath	2/lab	
Water bath	2/lab	Set at 45° to 55°C
Ruler	1/2S	Graduated in mm. To be provided by student.

EXERCISE 7
PURE CULTURE TECHNIQUES

Exercise 7 introduces the student to five important techniques in the isolation and maintenance of pure cultures. Parts A, B and C should be performed together and early in the course, preferably before staining techniques are done. Part D can be performed along with the previous three parts or in conjunction with those exercises in part IV. Part E, dealing with the use of selective and differential media, may be used as supportive material for those exercises dealing with the study of special groups of microorganisms (e.g., coliforms, staphylococci, and resident microbiota of the skin and throat).

OBJECTIVES:

A. Develop skills in streaking plates;

B. Develop skills in subculturing pure cultures;

C. Practice pipeting, especially with pipeting devices;

D. Understand the logic of using selective and/or differential media to isolate microorganisms from mixed populations.

REQUIREMENTS FOR EACH LABORATORY SECTION (FOR PARTS A, B AND C)

	QUANTITY	COMMENTS
Cultures		
Escherichia coli	4/lab	Cultured on TSB
Staphylococcus epidermis	4/lab	Cultured on TSB
Bacillus subtilis	4/lab	Cultured on TSB
Mixture of above bacteria	4/lab	Mix just before lab begins 1ml of each culture.
Media		
Trypticase soy agar deeps	3/S	In 18x150 tubes for pouring into plates. Prepoured plates may be used instead.
Trypticase soy broth	3/S	10 ml in 18x150 tubes.
Trypticase soy agar slants	3/S	In 18x150 tubes with a 2-3cm butt.
Petri plates	3/S	
Culture tube rack	1/S	
Boiling water bath	2/lab	
Incubator	1/lab	Set at 35°C.

REQUIREMENTS FOR EACH LABORATORY SECTION (PART D)

	QUANTITY	COMMENTS
Media		
Nutrient broth	3/S	10 ml in 18x150 tubes
Bottles of sterile water	1/S	90 ml in dilution bottles
Pipeting device	1/S	Propipets are recommended
Incubator	1/lab	Set at 35°C

REQUIREMENTS FOR EACH LABORATORY SECTION (FOR PART E)

	QUANTITY	COMMENTS
Cultures		
Escherichi coli	4/lab	Cultured on TSB
Staphylococcus epidermis	4/lab	Cultured on TSB
Staphylococcus aureus	4/lab	Cultured on TSB
Proteus vulgaris	4/lab	Cultured on TSB
Streptococcus lactis	4/lab	Cultured on TSB
S. mitis	4/lab	Cultured on TSB
S. faecalis	4/lab	Cultured on TSB
Media		
Mannitol-salts agar deeps	2/S	In 18x150 tubes for pouring into plates. Prepoured plates may be used instead.
EMB agar deeps	2/S	In 18x150 tubes for pouring into plates. Prepoured plates may be used instead.
Blood agar plates	3/S	TSA or BHI base with 5% sheep erythrocytes.
Petri plates	4/S	
Culture tube rack	1/S	
Boiling water bath	2/lab	
Incubator	1/lab	Set at 35°C

PART III
STAINING

EXERCISE 8
POSITIVE AND NEGATIVE STAINING

Exercise 8 introduces the student to two important techniques that facilitate the microscopic examination of microoranisms: preparation of smears and basic principles of staining microbial cells. It is recommended that both parts be performed in the same laboratory period but at least part A must be performed prior to introducing the student to the more sophisticated staining techniques (exercises 9 and 10). The student should master the preparation of smears before they attempt to perform the Gram or acid fast stain. The student should be reminded that preparing smears using heavy microbial suspensions will lead to unsatisfactory stains and should practice making smears from broth and agar cultures.

OBJECTIVES:

A.   Develop skills in preparing smears;

B.   Perform a positive stain;

C.   Perform a negative stain;

D. Compare the results of positive with negative staining procedures.

REQUIREMENTS FOR EACH LABORATORY SECTION

	QUANTITY	COMMENTS
Cultures		
Escherichi coli	4/lab	Cultured on TSA slants
Micrococcus luteus	4/lab	Cultured on TSB
Bacillus subtillis	4/lab	Cultured on TSB
Saccharomyces cerevisiae	4/lab	Cultured in Sabouraud broth
Stains		Available at all times in the lab bench.
Crystal violet	1 jar/2S	
Methylene blue	1 jar/2S	
Carbofuchsin	1 jar/2S	
Safranin	1 jar/2S	
Nigrosin	1 jar/2S	
India ink	1 jar/2S	
Glass slides	1 box/S	To be provided by student
Bibulous paper	1 pad/S	To be provided by student
Toothpicks	1 box/lab	Sterile in screw-capped tubes.

EXERCISE 9
DIFFERENTIAL STAINING

Exercise 9 introduces the student to the two most important staining techniques in the bacteriology laboratory: the Gram stain and the acid-fast stain. Both staining procedures may be performed during the same laboratory period. No other exercise should be scheduled during this time since the student must repeat these procedures several times in order to develop confidence in his/her skills. It is imperative that exponentially-growing cultures be provided to maximize the student's chances of obtaining good results. We recommend that the student master one of the staining procedures before performing the other. The student should be discouraged from making several smears and "mass process" the slides. Instead, he/she should prepare one smear, evaluate it, criticize it, and then stain the next smear.

OBJECTIVES:

A.   Develop skills in preparing smears;

B.   Perform a Gram stain;

C.   Perform an acid-fast stain;

D.   Evaluate what could happen if one of the steps in the procedure is either skipped or replaced with another.

REQUIREMENTS FOR EACH LABORATORY SECTION

	QUANTITY	COMMENTS
Cultures		
Escherichia coli	4/lab	Cultured on TSA slants
Staphylococcus epidermis	4/lab	Cultured on TSA slants
Bacillus subtilis	4/lab	Cultured on TSA slants
Neisseria subflava	4/lab	Cultured on TSA slants
Mycobacterium phlei	4/lab	Cultured on TSA slants
Stains*		Available at all times in the lab bench.
Exercise 9A		
Crystal violet	1 jar/2S	
Gram's iodine	1 jar/2S	
95% ethanol	1 jar/2S	
Safranin	1 jar/2S	
Exercise 9B		
Carbofuchsin	1 jar/2S	Ziehl-Neelsen modification
Methylene blue	1 jar/2S	
Acid-alcohol	1 jar/2S	
Glass slides	1 box/S	To be provided by student
Bibulous paper	1 pad/S	To be provided by student

*See Appendix in lab manual for preparation instructions.

EXERCISE 10
STRUCTURAL STAINING

Exercise 10 outlines common procedures to stain bacterial capsules, endospores, and flagella. We used the Maneval stain instead of India ink to stain the bacterial capsule because we have had considerable success with the procedure in our laboratory course for nonmajors. We emphasize, however, that the organisms are alive and that care must be exercised when handling or disposing of the stained preparations. We use the Schaffer-Fulton procedure for the endospore stain because it is simple to perform and yields excellent results. Similarly, the Gray's flagellar stain is very suitable for staining flagella, and if care is exercised while preparing the smears, excellent results are obtained. It is suggested that this exercise be the only new exercise performed during the period because it will require some time and effort on the part of the student to obtain acceptable results.

OBJECTIVES:

A. Observe some of the various structures of the bacterial cell;

B.   Review some of the strategies used in staining cells to show
     the desired characteristic;

C.   Evaluate what could happen if one of the steps in the procedure
     is either skipped or replaced with another.

REQUIREMENTS FOR EACH LABORATORY SECTION (FOR PARTS A, B AND C)

	QUANTITY	COMMENTS
Cultures		
Exercise 10A		
Klebsiella pneumoniae	4/lab	Cultured on TSA slants
Micrococcus luteus	4/lab	Cultured on TSA slants
Exercise 10B		
Bacilus subtilis	4/lab	Cultured on TSA slants
Clostridium sporogenes	4/lab	Cultured on TSA slants
C. butyricum	4/lab	Cultured on TSA slants
Exercise 10C		
Proteus vulgaris	4/lab	Cultured on TSA slants
Pseudomonas fluorescens	4/lab	Cultured on TSA slants
Stains*		Available at all times in the lab bench.
A. Congo red	1 jar/2S	
Maneval's stain (modified)	1 jar/2S	
B. Malachite green	1 jar/2S	
Safranin	1 jar/2S	
C. Gray's flagellar stain	4/class	Prepare fresh each time in 250ml flasks.
Acid-alcohol	1 jar/2S	
95% ethanol	1 jar/2S	
Glass slides	1 box/S	To be provided by student
Bibulous paper	1 pad/S	To be provided by student
Test tubes with 0.5ml saline	2/S	10x75 tubes with 0.85% NaCl

	QUANTITY	COMMENTS
Nitrocellulos filter, 0.22 µm	4/class	To filter Gray's stain
Pasteur pipets	2/S	

*See Appendix in lab manual for preparation instructions.

EXERCISE 11
DETECTION OF INCLUSIONS

Exercise 11 outlines common procedures to stain cellular inclusions such as volutin, carbohydrate, and PHB granules. This exercise can be conducted as part of a sequence of exercises aimed at demonstrating the various structural features of microorganisms or as part of the procedure for the identification of unknown bacteria. The student should be informed that microorganisms produce storage granules only under certain nutritional conditions, which are provided by the culture media used to culture them. The student should also be cautioned that the xylene is a potential carcinogen and should be handled with care. Students should be encouraged to wash their hands thoroughly after handling xylene and other chemical carcinogens.

OBJECTIVES:

A. Observe some of the various types of inclusions in microbial cells;

B.   Study why microorganisms store their excess nutrients in the forms of insoluble granules.

REQUIREMENTS FOR EACH LABORATORY SECTION

	QUANTITY	COMMENTS
Cultures		
A. Corynebacterium xerosis	4/lab	Cultured on tryptone-phosphate agar slants
Bacillus megaterium	4/lab	Cultured on tryptone-phosphate agar slants
B. Bacillus cereus	4/lab	Cultured on NA slants
Bacillus megaterium	4/lab	Cultured on NA slants
C. Bacillus subtilis	4/lab	Cultured on TSA slants
Saccharomyces cerevisiae	4/lab	Cultured on Sabouraud glucose agar slants
Stains/reagents*		Available at all times in the lab bench.
A. Toluidine blue	1 jar/2S	Or Loeffler's methylene blue aqueous, diluted 1:1000
Sulfuric acid solution	1 jar/2S	
Gram's iodine	1 jar/2S	
1% eosin Y	1 jar/2S	
B. Sudan black B	1 jar/2S	
Safranin	1 jar/2S	
Xylene	1 jar/2S	Careful! It is carcinogenic.
C. Gram's iodine	1 jar/2S	
Glass slides	1 box/S	To be provided by student
Bibulous paper	1 pad/S	To be provided by student

*See Appendix in lab manual for preparation instructions.

EXERCISE 12
MORPHOLOGICAL UNKNOWN

Exercise 12 can be used to test the student's understanding of the exercises in parts I and III and determine to which degree they have mastered the various staining and other microscopic techniques. If time allows, we strongly recommend that this exercise be performed right after the completion of exercise 11. Otherwise, this exercise can be incorported as part of exercise 28 or exercise 72.

OBJECTIVES:

A. Test the student's understanding of staining procedures and principles;

B. Determine the student's skills in preparing smears and staining microorganisms for microscopic examination.

REQUIREMENTS FOR EACH LABORATORY SECTION

	QUANTITY	COMMENTS
Microscope with micrometer	1/S	
Cultures of bacteria used in exercises 8-11	1/S	Label as unknowns. Different species can be handed out instead.
Stains and reagents	1/2S	Those used in exercises 8-11
Pasteur pipets	1/S	
Depression slide	1/S	
Vaseline	2 jars/lab	
Glass slides	1 box/S	To be provided by student
Bibulous paper	1 pad/S	To be provided by student
Lens paper	1 pad/S	To be provided by student

PART IV
ENUMERATION OF MICROORGANISMS

EXERCISE 13
DIRECT MICROSCOPIC COUNTS WITH A.
HEMOCYTOMETER

Exercise 13 illustrates the technique of counting cells in suspensions using a counting chamber. We use a yeast suspension because these cells are relatively large, nonmotile, and can be easily viewed at 400x magnication. Yeast cells with buds or adhering to each should be counted as one. This technique is not recommended for counting bacteria or motile cells.

OBJECTIVES:

A. Count a yeast suspension using a hemocytometer;

B. Recognize the advantages and shortcomings of direct microscopic methods.

REQUIREMENTS FOR EACH LABORATORY SECTION

	QUANTITY	COMMENTS
Microscope	1/S	

Cultures
  Saccharomyces
    cerevisiae
	1/4S	Grown on Sabouraud glucose broth

Hemocytometer	1/2S	
1 ml pipets	2/S	Sterilized
Pasteur pipets	2/S	
Rubber bulbs	1/S	
Sterile saline	2/S	In 13/100 screw-capped tubes

EXERCISE 14
VIABLE COUNTS

Exercise 14 illustrates two methods for performing viable counts of microbial suspensions: the pour plate method and the spread plate method. We use a sample of raw milk as the source of viable cells because raw milk is readily available, is easy to handle by students, is harmless, and yields predictable results. The student should be directed to exercise 7D to review the proper pipeting procedure and to the Appendix for a brief discussion of how to make dilutions. It should be emphasized that viable counts are minimal counts and that the formulation of the culture media can be modified (see exercise 7E) to count a selected group of microbes. Can be performed in pairs.

OBJECTIVES:

A. Perform a viable count on a sample of raw milk;

B. Practice pipeting and making dilutions;

C. Use a colony counter to count the colonies on/in the agar plates;

D. Recognize the advantages and shortcomings of viable counts.

REQUIREMENTS FOR EACH LABORATORY SECTION

	QUANTITY	COMMENTS
Raw milk	1/2S*	In 5 dram specimen vials
Plate count agar (PCA)+	6/2S	20 ml/tube in 18x150 tubes with closures
Petri plates	6/2S	Presterilized
Dilution blanks	8/2S	9 ml of sterile saline in 16x100 screw-capped tubes
1 ml pipets	10/2S	Sterilized
Sterile saline	2/S	In 13x100 screw-capped tubes
Glass spreaders	2/2S	Bent glass rods (like hockey sticks) immersed in jars containing 70% alcohol
Colony counter	4/lab	
Hand tallies	4/lab	

Water baths
  Boiling
  45° to 55°C
	
	4/lab
	2/lab
		
		To cool the agar

Incubator	1/lab	Set at 30° to 35°C

*Materials for exercise performed in pairs.
+PCA is Difco's trade name for tryptone-phosphate agar. BBL's name is Standard Methods Agar.

EXERCISE 15
OPTICAL DENSITY MEASUREMENTS

Exercise 15 provides the student with the opportunity of making optical density measurements of a bacterial population using a colorimeter. The student will also be asked to construct a calibration curve so that he/she can estimate the number of microorganisms in a suspension from O.D. readings. This exercise should be performed in pairs in order to economize in supplies, equipment, and time. This exercise should be performed after exercise 14 so that the student is more skilled in making dilutions and carrying out viable counts. It should be emphasized that the calibration curve obtained is only useful when measuring population sizes of <u>Escherichia</u> <u>coli</u> growing in glucose-salts broth.

OBJECTIVES:

A.   Perform a viable count on a bacterial suspension;

B.   Determine the O.D. of a bacterial suspension using a colorimeter;

C.   Construct a calibration curve of O.D. vs viable counts (log);

D.   Practice pipeting and making dilutions;

E.   Use a colony counter to count the colonies on/in the agar
      plates;

F.   Recognize the advantages and shortcomings of turbidimetric
      counts.

REQUIREMENTS FOR EACH LABORATORY SECTION

	QUANTITY	COMMENTS
Cultures Escherichia coli	1/2S*	Overnight cultures on glucose-salts medium·
Colorimeter with cuvettes	4/lab	Spectronic 20 or other suitable colorimeter
Plate count agar (PCA)+	6/2S	20 ml/tube in 18x150 tubes with closures
Petri plates	6/2S	Presterilized
Dilution blanks	4/2S	9 ml of sterile saline in 16x100 screw-capped tubes
Glucose-salts broth	6/2S	In 16x100 screw-capped tubes
1 ml pipets	4/2S	Sterilized
10 ml pipets	6/2S	Sterilized
Colony counter	4/lab	
Hand tallies	4/lab	

Water baths
  Boiling
  45° to 55°C
	4/lab
	2/lab
		To cool the agar

Incubator	1/lab	Set at 30° to 35°C

*Materials for exercise performed in pairs.
+PCA is Difco's trade name for tryptone-phosphate agar. BBL's name
  is Standard Methods Agar.
·See Appendix for formulation.

PART V
ENVIRONMENTAL FACTORS
AFFECTING GROWTH

EXERCISE 16
EFFECT OF TEMPERATURE
ON MICROBIAL GROWTH

Exercise 16 illustrates the effect of environmental temperature on microbial growth. In this exercise, the student will be asked to incubate mesophiles, thermophiles, and psychrophiles at various temperatures, ranging from 4°C to 60°C and determine the amount of growth as evidenced by turbidity, sediment formation, or pellicle formation. The exercise requires that six incubators be used. If this is impossible, we recommend that the number of incubation temperatures be reduced to 10°C, 35°C, and 55° C. The cold environment can be achieved by using a refrigerator with the temperature control turned up to minimum refrigeration, the other two environmental temperatures can be achieved using conventional incubators. This exercise can be performed in pairs, or set up as a demonstration, along with the other exercises in this part (V).

OBJECTIVES:

A.   Demonstrate the effect of temperature on microbial growth;

B.   Become acquainted wth the various groups of microorganisms classified on their growth temperature ranges.

REQUIREMENTS FOR EACH LABORATORY SECTION

	QUANTITY	COMMENTS
Cultures		
Pseudomonas fluorescens	4/lab	Overnight culture on TSB
Escherichia coli	4/lab	Overnight culture on TSB
Mycobacterium phlei	4/lab	Overnight culture on TSB
Bacillus stearo-thermophilus	4/lab	Overnight culture on TSB
Media		
Trypticase soy broth	6/2S	10 ml in 18x150 culture tubes with closures.
Incubators		
4°C incubator	1/lab	A refrigerator is suitable
18°C incubator	1/lab	A B.O.D. incubator is needed
37°C incubator	1/lab	
45°C incubator	1/lab	
55°C incubator	1/lab	
60°C incubator	1/lab	

EXERCISE 17
EFFECT OF pH ON MICROBIAL GROWTH

Exercise 17 illustrates the effect of environmental pH on microbial growth. In this exercise, the student will be asked to culture neutrophiles, acidophiles, and alkalinophiles on culture media of various pH values, ranging from 3 to 12, and determine the amount of growth that occurs. The various pH can be obtained by adding 1N HCl or 1N KOH to trypticase soy broth during preparation. This exercise can be performed in pairs, or set up as a demonstration, along with the other exercises in this part (V).

OBJECTIVES:
A.   Demonstrate the effect of pH on microbial growth;
B.   Become acquainted with the various groups of microorganisms classified on their growth pH ranges.

REQUIREMENTS FOR EACH LABORATORY SECTION

	QUANTITY	COMMENTS
Cultures		
Pseudomonas fluorescens	4/lab	Overnight culture on TSB
Escherichia coli	4/lab	Overnight culture on TSB
Lactobacillus bulgaricus	4/lab	Overnight culture on TSB
Saccharomyces sp.	4/lab	2-day culture on Sab slants*
Penicillium sp.	4/lab	7-day culture on Sab slants
Rhizopus sp.	4/lab	7-day culture on Sab slants
Media		
Trypticase soy broth (pH values 3-12)	6/2S	10 ml in 18x150 culture tubes with closures. pH can be adjusted by adding 1N HCl or 1N KOH to the TSB tubes.
Incubator	1/lab	Set at 25° to 30°C

*Sab = Sabouraud dextrose agar

EXERCISE 18
EFFECT OF OSMOTIC PRESSURE
ON MICROBIAL GROWTH

Exercise 18 demonstrates the effect of environmental osmotic pressure on microbial growth. In this exercise, the student will be asked to culture various bacteria and fungi on culture media of varying osmotic pressures, using salt and sucrose, and determine the amount of growth that occurs. The various osmotic pressures can be acheived by adding NaCl or sucrose to nutrient agar during their preparation. This exercise can be performed in pairs, or set up as a demonstration, along with the other exercises in this part (V).

OBJECTIVES:

A. Demonstrate the effect of environmental osmotic pressure on microbial growth;

B. Become acquainted with the various groups of microorganisms classified on their ability to grow in environments of various osmotic pressures.

REQUIREMENTS FOR EACH LABORATORY SECTION

	QUANTITY	COMMENTS
Cultures		
Escherichia coli	4/lab	Overnight culture on NB*
Staphylocossus sp.	4/lab	Overnight culture on NB
Saccharomyces sp.	4/lab	2-day culture on Sab slants
Penicillium sp.	4/lab	7-day culture on Sab slants
Rhizopus sp.	4/lab	7-day culture on Sab slants
Media		
A. Nutrient agar (with NaCl)	7/2S	20 ml in 18x150 culture tubes with closures. Adjust salt concentration by adding sufficient NaCl to make final NaCl concentrations of: 0.5%, 1%, 3%, 5%, 10%, 15%, and 20%.
B. Nutrient agar (with sucrose)	7/2S	20 ml in 18x150 culture tubes with closures. Adjust sugar concentration by adding sufficient sucrose to make final sugar concentrations of: 1%, 5%, 10%, 20%, 40%, and 60%.
Incubator	1/lab	Room temperature (~18° C)

*NB = nutrient broth

EXERCISE 19
EFFECT OF OXYGEN
ON MICROBIAL GROWTH

Exercise 19 demonstrates the effect of environmental oxygen on microbial growth. In this exercise, the student will be asked to incubate various bacteria under aerobic, anaerobic, and microaerophilic environments and determine the amount of growth that occurs. Anaerobic environments will be achieved using a GasPak or pyrogallic acid and NaOH. Microaerophilic environments will be achieved using a candle jar. This exercise can be performed in pairs, or set up as a demonstration, along with the other exercises in this part (V).

OBJECTIVES:

A. Demonstrate the effect of molecular oxygen on microbial growth;

B. Become acquainted with the various groups of microoranisms classified on their ability to grow in the presence of oxygen.

REQUIREMENTS FOR EACH LABORATORY SECTION

	QUANTITY	COMMENTS
Cultures		For exercises 19A-19D
Escherichia coli	4/lab	Overnight culture on TSB*
Clostridium perfringens	4/lab	48-hour culture on TSB
Micrococcus luteus	4/lab	Overnight culture on TSB
Neisseria sicca	4/lab	Overnight culture on TSB
Streptococcus lactis	4/lab	48-hour culture on TSB
Exercise 19A		
Plate count agar+	5/2S	20 ml in 18x150 culture tubes with closures
GasPak jars	4/lab	With fresh catalysts
GasPak envelopes	4/lab	
Anaerobic indicator	4/lab	
10 ml pipets	4/lab	
Distilled water	50ml/lab	In 100 ml beaker
Petri plates	10/2S	
Exercise 19B		
Plate count agar	10/2S	20 ml in 18x150 culture tubes with closures.
1 ml pipets	5/2S	
Exercise 19C		
Plate count agar	10/S	20 ml, slanted, in 18x150 culture tubes with closures.
Pyrogallic acid	100g	Provide dispensing spatula
4% NaOH	100ml	In bottle with dropper
Scissors	1/4S	
Rubber stoppers	5/2S	Size 2
Exercise 19D		
Plate count agar	10/2S	20 ml in 18x150 culture tubes with closures.
Candle jar	4/lab	With candle and matches

REQUIREMENTS FOR EACH LABORATORY SECTION

	QUANTITY	COMMENTS
Petri plates	10/S	
Incubator	1/lab	Set at 35°C

*TSB = trypticase soy broth
+PCA is Difco's trade name for tryptone-phosphate agar. BBL markets
 this medium under the name of Standard Methods Agar.

EXERCISE 20
EFFECT OF LIGHT
ON MICROBIAL GROWTH

Exercise 20 demonstrates the effect of light on microbial growth. In this exercise, the student will be asked to prepare Winogradsky columns and incubate them in the presence of light or in the dark. 3M salt will be added to the aqueous part of some of the columns to select for phototrophic halophiles. This experiment does not require much time to conduct and yields very satisfactory results. We have included the requirements for each group of four students although the experiment can be set up as a demonstration. Sunlight or incandescent light are suitable light sources but fluorescent light is not.

OBJECTIVES:

A.   Demonstrate the effect of light on microbial growth;

B.   Become acquainted with the various groups of microorganisms based on their light requirements.

REQUIREMENTS FOR EACH LABORATORY SECTION

	QUANTITY	COMMENTS
Rich soil or mud	4 gallon	From garden, lake shore or estuary.
Pond water	2 gallon	
1L graduated cylinder	4/4S	
Incandescent light source	4/lab	
Chemicals NaCl	500 ml	3M solution
K_2HPO_4	100 g	
KH_2PO_4	100 g	
Cellulose	100 g	Shredded filter paper or paper towels can be used instead of cellulose (enough for 1 handful/column).
Aluminum foil	1 roll	

PART VI
CONTROL OF MICROBIAL GROWTH

EXERCISE 21
PHYSICAL METHODS OF CONTROL

Exercise 21 illustrates the effectiveness of various physical control measures. We use handwashing (21A) to illustrate to students that handwashing is indeed effective in removing large quantities of microorganisms from the skin. Experiments 21B and 21D demonstrate that microbiocidal agents vary in effectiveness depending upon the microorganism involved and that microbial populations do not die immediately upon exposure to the killing agent, but rather only a portion of the population dies during any given time interval. Exercise 21C shows that fluids are sterilized using filters because the filter does not allow microbes to pass and that this is a function of the pore size.

OBJECTIVES:

A. Demonstrate experimentally the effect of physical agents on the reproduction of microoranisms;

B. Show the effectiveness of handwashing in removing microorganisms from the skin;

C. Demonstrate the microbiocidal activity of boiling and ultraviolet light;

D. Show the relationship between pore size of membrane filters and the ability of these filters to sterilize bacterial suspensions.

REQUIREMENTS FOR EACH LABORATORY SECTION

	QUANTITY	COMMENTS
Exercise 21A:		
Soil suspension	1/5S	1/10 dilution in 18/150 tubes
Soap	4 bars	
Cotton swabs	2/S	
Saline	2/S	2 ml in 10x75 tubes
Nutrient agar	2/S	20 ml in 18x150 tubes
Petri plates	2/S	
Exercise 21B:		
Cultures _Escherichia coli_	1/4S	24-hour culture in TSB (35°C)
Bacillus stearo-thermophilus	1/4S	24-hour culture in TSB (35°C)
Trypticase soy broth (TSB)	10/4S	10 ml in 18x150 tubes
Boiling water bath	1/4S	
Pipets, 1 ml	10/4S	
Incubators	2/lab	Set at 35°C and at 55°C
Exercise 21C:		
Cultures _Escherichia coli_	1/4S	Overnight cultures in TSB

REQUIREMENTS FOR EACH LABORATORY SECTION

	QUANTITY	COMMENTS
Filtering apparatus	1/4S	
Membrane filters	5/4S	One of each of the following pore diameters: 1 m, 0.5 m, 0.3 m, 0.2 m, and 0.1 m.
Trypticase soy broth	5/4S	10 ml in 18x150 tubes
1 ml pipets	5/4S	
Incubator	1/lab	Set at 35°C

Exercise 21D:

Cultures		
Serratia marcescens	4/lab	Overnight cultures in TSB
Bacillus stearo-thermophilus	4/lab	Overnight cultures in TSB
Escherichia coli	4/lab	Overnight cultures in TSB
Ultraviolet light source	1/4S	In protective boxes
Protective eye glasses	1/S	
Stop watch		
Plate count agar	24/4S	
Petri plates	24/4S	
Glass spreaders	8/4S	In alcohol jars
Incubators	2/lab	Set at 18°C and at 55°C

EXERCISE 22
CHEMICAL METHODS OF CONTROL

Exercise 22 demonstrates the ability of chemical control agents to inhibit or kill microbial populations. The experiments in this exercise are more involved than many of those that appear in other laboratory manuals but we feel that the additional effort required in preparing for and performing the experiments will be well worth it if one considers the added experience, practice and knowledge that the students would gain. In some cases, we have asked the student to prepare certain solutions of antimicrobials without telling him/her exactly how to do it because it will force the student to analyze the exercise in more detail and gain needed experience in performing dilutions. To reduce the volume of work each student has to perform, we suggest that the experiments be performed in groups of four, each group performing one part of the exercise and then sharing the results with the rest of the class. For example, in experiment 22A each group tests one organism and a particular incubation temperature (ice bath, room temperature or

45°C) and then shares their results with the rest of the class. Additionally, only one or two of the experiments may be performed during a given laboratory period, rather than all four experiments.

OBJECTIVES:

A.   Demonstrate experimentally the effect of chemical agents on the reproduction of microoranisms;

B.   Calculate the phenol coefficient of a disinfectant;

C.   Demonstrate experimentally the nature of a broad spectrum and a narrow spectrum antibiotic;

D.   Demonstrate experimentally the concept of selective toxicity.

REQUIREMENTS FOR EACH LABORATORY SECTION

	QUANTITY	COMMENTS
Exercise 22A:		
Cultures		
Staphylococcus aureus	1/4S	Overnight cultures in TSB
Bacillus subtilis	1/4S	Overnight cultures in TSB
3% Hydrogen peroxide	100 ml	
70% isopropyl alcohol	100 ml	
Lysol cleaner	1 bottle	
Lysol disinfectant	1 bottle	
Trypticase soy broth	64/4S	Should be dispensed in 5 ml aliquots in 16x100mm tubes
5 ml pipets	20/4S	
1 ml pipets	1/4S	
Ice bath	1/4S	

REQUIREMENTS FOR EACH LABORATORY SECTION

	QUANTITY	COMMENTS
Water bath at room temperature	1/4S	
Water bath at 45°C	2/lab	
Exercise 22B:		
Cultures Staphylococcus aureus	1/4S	24-hour culture in TSB (35°C)
Lysol disinfectant	100 ml	
Trypticase soy broth (TSB)	39/4S	5 ml in 16x100 tubes
Dilution blanks	15/4S	5 ml of sterile distilled water in 16x100 tubes.
Incubator	1/lab	Set at 35°C
Exercise 22C:		
Cultures Escherichia coli	1/4S	Overnight cultures in TSB
Trypticase soy broth	30/4S	5 ml in 16x100 tubes
TSB with 0.08M sulfanilamide	1/4S	5 ml in 16x100 tubes
TSB+400 g/ml streptomycin	1/4S	5 ml in 16x100 tubes
TSB+200 units/ml penicillin	1/4S	5 ml in 16x100 tubes
1 ml pipets	2/4S	
5 ml pipets	15/4S	
Incubator	1/lab	Set at 35°C
Exercise 22D: Cultures		
Staphylococcus aureus	1/4S	48-hour cultures in TSB
Saccharomyces cerevisiae	1/4S	48-hour cultures in TSB

REQUIREMENTS FOR EACH LABORATORY SECTION

	QUANTITY	COMMENTS

Media
  TSB with 0.02M
  sulfanilamide
	2/4S	5 ml in 16x100 tubes

TSB+100 units/ml penicillin	2/4S	5 ml in 16x100 tubes
TSB+10 units/ml mycostatin	2/4S	5 ml in 16x100 tubes
1 ml pipets	2/4S	
Incubator	1/lab	Set at 35°C

PART VII
METABOLIC ACTIVITIES
OF MICROORGANISMS

EXERCISE 23
HYDROLYSIS OF LARGE
EXTRACELLULAR MOLECULES

Exercise 23 is the first of six exercises demonstrating the metabolic activities of microorganisms. This particular exercise deals with those biological polymers that are found in the extracellular environment that can be used as sources of nutrients by microorganisms. All or part of the exercise can be performed in a single period. We suggest that exercises 23-27 be performed before exercise 28. An alternative method would be to provide each student with an unknown culture (exercise 28) and then have him/her gradually perform all those exercises outlined in the report form for exercise 28. We recommend, however, that each student or group of students perform controls for each exercise if the latter approach is desired.

OBJECTIVES:

A. Demonstrate experimentally the ability of microorganisms to produce exoenzymes to digest extracellular polymers;

B.   Illustrate that the ability to utilize extracellular molecules
     is a function of the genetic makeup of the organism.

REQUIREMENTS FOR EACH LABORATORY SECTION

	QUANTITY	COMMENTS
Exercise 23A:		
Cultures		
Escherichia coli	1/4S	Overnight cultures in TSB
Bacillus subtilis	1/4S	Overnight cultures in TSB
Media		
Starch agar	2/S	20 ml in 18x150mm tubes
Petri dishes	2/S	
Gram's iodine	1/2S	Available in lab bench
Boiling water bath	4/lab	
Water bath	1/lab	Set at 55°C
Incubator	1/lab	Set at 35°C
Exercise 23B:		
Cultures		
Escherichia coli	1/4S	Overnight cultures in TSB
Bacillus subtilis	1/4S	Overnight cultures in TSB
Media		
Plate count agar	2/S	20 ml in 18x150mm tubes
Sterile skim milk	2/S	2 ml in 13x100mm tubes
Petri dishes	2/S	
Boiling water bath	4/lab	
Water bath	1/lab	Set at 55°C
Incubator	1/lab	Set at 35°C
Exercise 23C:		
Cultures		
Escherichia coli	1/4S	Overnight cultures in TSB
Serratia marcescens	1/4S	Overnight cultures in TSB

REQUIREMENTS FOR EACH LABORATORY SECTION

	QUANTITY	COMMENTS
Media		
DNAse agar plates	2/S	20 ml in 100x20mm plates
Incubator	1/lab	set at 35°C
Exercise 23D:		
Cultures		
Proteus mirabilis	1/4S	Overnight cultures in TSB
Staphylococcus epidermidis	1/4S	Overnight cultures in TSB
Media		
Spirit blue agar	2/S	With Bacto-lipase reagent or other lipid (see manual). In 20 ml in 18x150mm tubes.
Incubator	1/lab	Set at 35°C

EXERCISE 24
FERMENTATION OF CARBOHYDRATES

Exercise 24 illustrates how microorganisms use and modify carbohydrate-containing culture media. It also illustrates that the ability of microorganisms to ferment carbohydrates and the waste products they produce can be used as criteria to identify them. Brom cresol purple (BCP) or phenol red are equally suitable to detect changes occurring in the media; we use BCP because it is our preference. The methyl red and Voges-Proskauer tests are discussed individually, each with its own "materials" list. Both these tests can be performed concurrently by inoculating a tube of MR-VP broth with the test organism. After the incubation period, the culture is split equally between two tubes and treated with reagents as appropriate.

OBJECTIVES:

A. Demonstrate experimentally the ability of microorganisms to ferment carbohydrates;

B.   Illustrate that the ability to utilize mono and   disaccharides
     and   the   waste products of fermentation are functions   of   the
     genetic makeup of the organism.

REQUIREMENTS FOR EACH LABORATORY SECTION

	QUANTITY	COMMENTS
Exercise 24A:		
Cultures		
Escherichia coli	1/4S	Overnight cultures in TSB
Staphylococcus aureus	1/4S	Overnight cultures in TSB
Micrococcus luteus	1/4S	Overnight cultures in TSB
Saccharomyces cerevisiae	1/4S	48-hour cultures in TSB
Media*		20 ml in 18x150mm tubes with 10x74 Durham tubes.
BCP-glucose broth	2/S	
BCP-sucrose broth	2/S	
BCP-lactose broth	2/S	
BCP-maltose broth	2/S	
BCP-mannitol broth	2/S	
Incubator	1/lab	Set at 35°C
Exercises 24B and 24C:		
Cultures		
Escherichia coli	1/4S	Overnight cultures in TSB
Enterobacter aerogenes	1/4S	Overnight cultures in TSB
Media		
MR-VP broth	2/S	20 ml in 18x150 tubes
Reagents		See Appendix and Manual for preparation instructions.
Methyl Red	4/lab	In bottles with dropper
40% KOH	4/lab	In bottles with dropper
Alpha-naphthol	4/lab	In bottles with dropper
Incubator	1/lab	Set at 35°C

*See Appendix in lab manual.

EXERCISE 25
RESPIRATION OF CARBOHYDRATES

Exercise 25 demonstrates some aspects of respiration by microorganisms. All three exercises can be performed in one laboratory period. Each student should perform all the experiments since they are simple to do and require very little time to do them. We suggest that each student streak a plate with each of the bacteria in exercises 25A and 25B the period before the exercises are to be done so that they have well-developed, isolated colonies with which to perform the oxidase and catalase tests.

OBJECTIVES:

A. Demonstrate experimentally the ability of microorganisms to respire carbohydrates;

B. Compare and contrast aerobic and anaerobic respiration;

C. Compare respiration with fermentation;

D. Perform and interpret the catalase, oxidase, and nitrate reduction tests.

REQUIREMENTS FOR EACH LABORATORY SECTION

	QUANTITY	COMMENTS
Exercise 25A:		
Cultures		
Escherichia coli	1/4S	Overnight cultures in TSB
Pseudomonas aeruginosa	1/4S	Overnight cultures in TSB
Media		
Nutrient agar	2/S	20 ml x 18x150 tubes for pouring plates. These may be handed out the period before, along with the cultures, so that the student can streak the plates then and have colonies to test.
Oxidase reagent	1/4S	In bottles with droppers. See manual and appendix for formulation.
Pasteur pipets with bulbs	1/S	
Incubator	1/lab	Set at 35°C
Exercise 25B:		
Cultures		
Streptococcus faecalis	1/4S	48-hour cultures in TSB
Staphylococcus epidermidis	1/4S	48-hour cultures in TSB
Media		
Nutrient agar	2/S	20 ml in 18x150 tubes for pouring plates. These may be handed out the period before, along with the cultures, so that the student can streak the plates and then have colonies to test.
Catalase reagent	1/4S	In bottles with droppers. 3% aqueous hydrogen peroxide.

REQUIREMENTS FOR EACH LABORATORY SECTION

	QUANTITY	COMMENTS
Pasteur pipets with bulbs	1/S	
Incubator	1/lab	Set at 35°C
Exercise 25C:		
Cultures		
Escherichia coli	1/4S	Overnight cultures in TSB
Pseudomonas aeruginosa	1/4S	Overnight cultures in TSB
Media		
Nitrate broth	2/S	20 ml in 18x150 tubes with Durham tubes.
Reagents		
Sulfanilic acid	1/4S	In bottle with dropper. See appendix for formulation.
Alpha-naphthylamine	1/4S	In bottle with dropper. See appendix for formulation.
Sulfamic acid		This can be used instead of the two reagents above. The formulation is also given in the appendix. Results are + when gas bubbles appear.
Zn dust	1g	Provide with spatula.
Incubator	1/lab	Set at 35°C

EXERCISE 26
UTILIZATION OF AMINO ACIDS

Exercise 26 illustrates some aspects of amino acid utilization by microorganisms. All four parts of the exercise can be performed in one laboratory period. Each student should perform all four since they are simple to do and require very little culture media. This exercise contains important biochemical tests that aid in the identification of unknowns, particularly those in the family Enterobacteriaceae. The organisms listed in this exercise are control organisms of which >90% of the strains are positive for the reaction (positive controls) or negative for the reaction (negative controls).

OBJECTIVES:

A. Demonstrate experimentally the ability of microorganisms to utilize and modify amino acids;

B. Perform and interpret the indole production, hydrogen sulfide production, lysine decarboxylation, and phenylalanime deamination tests.

REQUIREMENTS FOR EACH LABORATORY SECTION

	QUANTITY	COMMENTS
Exercises 26A and 26B:		
Cultures		
Escherichia coli	1/4S	Overnight cultures in TSB
Enterobacter aerogenes	1/4S	Overnight cultures in TSB
Proteus vulgaris	1/4S	Overnight cultures in TSB
Media		
SIM agar deeps	3/S	10-20 ml in 18x150 tubes
Reagents		
Kovak's reagent	1/4S	In bottles with droppers. See manual and appendix for formulation.
Pasteur pipets with bulbs	1/S	
Incubator	1/lab	Set at 35°C
Exercise 26C:		
Cultures		
Enterobacter aerogenes	1/4S	Overnight cultures in TSB
Citrobacter freundii	1/4S	Overnight cultures in TSB
Media		
Lysine decarboxylase broth with lysine	2/S	5 ml in 13x100 tubes. See appendix for formulation of Moller's formulation.
Lysine decarboxylase broth without lysine	2/S	5 ml in 13x100 tubes. See appendix for formulation of Moller's formulation.
Mineral oil, sterile	1/4S	In bottles with droppers.
Pasteur pipets with bulbs	1/S	
Incubator	1/lab	Set at 35°C
Exercise 26D:		
Cultures		
Escherichia coli	1/4S	Overnight cultures in TSB

REQUIREMENTS FOR EACH LABORATORY SECTION

	QUANTITY	COMMENTS
<u>Proteus</u> <u>vulgaris</u>	1/4S	Overnight cultures in TSB

Media
  Phenylalanine agar
    slants
	2/S	20 ml in 18x150 tubes with a 1 " butt.

Reagents
  10% aqueous ferric
    chloride
	1/4S	In bottle with dropper. See appendix for formulation.

Pasteur pipets	1/S	
Incubator	1/lab	Set at 35°C

EXERCISE 27
UTILIZATION OF
CITRATE, GELATIN, AND UREA

Exercise 27 detects microbial activities that are carried out by microorganisms and are commonly used in the identification of unknown bacteria. The citrate test, one of the procedures in the IMViC test for enterics, measures the ability of bacteria to utilize citrate as the sole source of carbon. In the introduction, it might be interesting to the student to know that the medium is a chemically-defined medium rather than a complex one. The other two procedures can be used to illustrate microbial activities in nature and elucidate the role of bacteria as decomposers and modifiers of organic matter. Each student should perform the three procedures outlined in exercise 27 because they provide important clues in the identification of unknown bacteria.

OBJECTIVES:

A. Demonstrate experimentally the ability of microorganisms to utilize and modify various chemical substrates found in their environment;

71

B.   Perform   and   interpret   the   citrate   utilization,   gelatin
     hydrolysis, and the urea hydrolysis tests.

C.   Illustrate   that   microoranisms   function   as   demomposers   by
     altering organic matter in their environment as they use it   as
     sources of nutrients.

REQUIREMENTS FOR EACH LABORATORY SECTION

	QUANTITY	COMMENTS
Exercise 27A:		
Cultures		
Escherichia coli	1/4S	Overnight cultures in TSB
Enterobacter aerogenes	1/4S	Overnight cultures in TSB
Media		
Simmon's citrate agar	2/S	10-20 ml in 18x150 tubes, slanted, with a 1" butt.
Incubator	1/lab	Set at 35°C
Exercise 27B:		
Cultures		
Escherichia coli	1/4S	Overnight cultures in TSB
Bacillus subtilis	1/4S	Overnight cultures in TSB
Media		
Nutrient gelatin deeps	2/S	10-20 ml in 18x150 tubes
Ice water bath	4/lab	Or refrigerator
Incubator	1/lab	Set at 35°C
Exercise 27C:		
Cultures		
Escherichia coli	1/4S	Overnight cultures in TSB
Proteus vulgaris	1/4S	Overnight cultures in TSB
Media		
Christensen's urea agar	2/S	20 ml in 18x150 tubes, slanted, with a 1" butt.

REQUIREMENTS FOR EACH LABORATORY SECTION

	QUANTITY	COMMENTS
Reagents		
Incubator	1/lab	Set at 35°C

EXERCISE 28
IDENTIFICATION OF
AN UNKNOWN BACTERIUM

Exercise 28 tests the student's ability to identify an unknown bacterium by applying the knowledge and skills they have acquired during the performance of the various exercises in this and other parts of the lab manual. We recommend that the student receive as a first unknown (if time permits more than one unknown) a pure culture. Subsequent unknowns may consist of two or more organisms freshly mixed, which the student would have to isolate and then test. Two approaches to this exercise can be used: (a) hand out unknown after the student has performed all tests in part VII; or (b) hand out the unknown after the student has learned how to streak and subculture organisms, then allow him/her to perform all other tests on the unknown (using the organisms given in the exercises as positive and negative controls).

OBJECTIVES:

A.   Identify an unknown bacterium;

B.   Test the student's retention and understanding of the tests previously performed;

C.   Develop the student's confidence to carry out microbiological techniques properly.

REQUIREMENTS FOR EACH LABORATORY SECTION

	QUANTITY	COMMENTS
Cultures	1/S	May use the organisms employed in other exercises.
Media	1/S	All those used in previous exercises should be available.
Reagents	1/4S	Those used in previous exercises.
Incubators		As needed.

PART VIII
SURVEY OF
EUKARYOTIC MICROORGANISMS

EXERCISE 29
THE YEASTS AND MOLDS

Exercise 29 is aimed at introducing the students to some of the characteristics of the fungi. It involves the examination of prepared slides to get an idea of the morphological diversity that exists among the fungi and the cultivation of fungi. Emphasis will be placed on their morphological characteristics rather than on their physiology.

OBJECTIVES:

A. Observe various species and groups of fungi to acquire an appreciation of the diversity that exists in the kingdom, Fungi.

B. Cultivate the fungi to gain knowledge of the way in which fungi reproduce in the laboratory;

C. Learn some techniques for the examination of fungi in cultures;

D. Demonstrate the sexual characteristics of the Zygomycetes.

REQUIREMENTS FOR EACH LABORATORY SECTION

	QUANTITY	COMMENTS
Exercise 29A:		
Cultures		
Aspergillus	1/4S	7-day cultures on Sab slants
Penicillium	1/4S	7-day cultures on Sab slants
Alternaria	1/4S	7-day cultures on Sab slants
Geotrichum	1/4S	7-day cultures on Sab slants
Saccharomyces	1/4S	7-day cultures on Sab slants
Rhizopus	1/4S	7-day cultures on Sab slants
Prepared slides		Obtain from Carolina Biological Supply Co.
Morchella	1/4S	
Puccinia	1/4S	
Agaricus	1/4S	
Media		
Sabouraud agar	6/2S	20 ml in 18x150 tubes, slanted, with a 1" butt.
Lactophenol cotton blue	1/4S	In bottles with dropper.
Slides and cover-slips	6/S	To be provided by student.
Inoculating needles	1/S	With tip bent at 90° angle.
Dissecting needles	1/S	
Sand jar with disinfectant	1/2S	To clean needle tip before flaming the needle. Using a 50 ml wide mouth jar add a 50:50 mixture of sand and disinfectant to the jar.
Incubator	1/lab	Set at 28°C
Exercise 29B:		
Cultures		
Aspergillus	1/4S	7-day cultures on Sab slants
Penicillium	1/4S	7-day cultures on Sab slants
Alternaria	1/4S	7-day cultures on Sab slants
Media		
Cornmeal agar deeps	3/S	20 ml in 18x150 tubes
Petri dishes	3/S	

REQUIREMENTS FOR EACH LABORATORY SECTION

	QUANTITY	COMMENTS
Flat blade spatula	1/S	In alcohol jars to cut agar block.
Sterile coverslips	3/S	Coverslips should be placed in alcohol jars for about 30 minutes before use.
Slides	3/S	To be provided by student.

Reagents
  Lactophenol cotton
  blue
	1/4S	In bottles with dropper

Sand jar with disinfectant	1/2S	To clean needle tip before flaming the needle. Using a 50 ml wide mouth jar add a 50:50 mixture of sand and disinfectant to the jar.
Incubator	1/lab	Set at 28°C
Exercise 29C:		
Cultures		May purchase _Rhizopus_ zygospore kit from Caroline Biological (#15-5827)
Rhizopus _stolonifer_ (+)	1/4S	4-day cultures in Sab slants
Rhizopus _stolonife_ (-)	1/4S	4-day cultures in Sab slants

Media
  Sabouraud glucose
  agar
	2/S	20 ml in 18x150 tubes

Reagents
  Lactophenol cotton
  blue
	1/4S	In bottles with dropper.

Slides and cover- slips	1/S	To be provided by student.
Sand jar with disinfectant	1/2S	See exercise 29A or 29B
Incubator	1/lab	Set at 28°C

EXERCISE 30
THE PROTOZOA

Exercise 30 introduces the student to the major groups of protozoa. The student is asked to examine prepared slides, living cultures, and pond water. Protoslo should be used to examine the living cultures and the pond water because it slows down the motility of these organisms and make it easier to examine them microscopically. The prepared slides include some of the most common human pathogens so that the student becomes familiar with the parasites, which would likely be studied in the lecture portion of the course. This would make the study of the diseases caused by these agents more interesting.

OBJECTIVES:

A. Observe various groups of protozoa in order to acquire an appreciation of the characteristics of the group as well as the species diversity;

B. Examine pond water to determine which are the predominant groups of protozoans there.

REQUIREMENTS FOR EACH LABORATORY SECTION

	QUANTITY	COMMENTS
Microscope	1/S	
Cultures		Obtain from Carolina Biological Supply Co.
Paramecium	1/4S	Catalog #L 2A
Amoeba	1/4S	Catalog #L 1
Euglena	1/4S	Catalog #L 3
Or a mixture of protozoa	1/4S	Catalog #L 52J
Prepared slides		Obtain from Carolina Biological Supply Co.
Giardia intestinalis	1/4S	PS 210 and PS 220
Entamoeba histolytica	1/4S	PS 150 and PS 160
Plasmodium vivas	1/4S	PS 600
Trypanosoma lewisi	1/4S	PS 285
Balantidium coli	1/4S	PS 1100
Pond water	1/4S	Students may be asked to bring a sample to class.
Slides and cover-slips	3/S	To be provided by student.
Pasteur pipets	6/S	
Protoslo	1/4S	Carolina Biological Supply Co. #88-5141

EXERCISE 31
THE ALGAE

Exercise 31 introduces the student to some of the major groups of algae. The student will study this diverse group of organisms by examining living cultures and pond water. This exercise can be performed along with exercise 30. This way, the same samples of pond water can be used for both exercises 30 and 31.

OBJECTIVES:

A. Observe various groups of unicellular algae in order to acquire an appreciation of the characteristics of the group as well as the species diversity;

B. Examine pond water to see which are the predominant groups of unicellular algae there.

REQUIREMENTS FOR EACH LABORATORY SECTION

	QUANTITY	COMMENTS
Microscope	1/S	
Cultures		Obtain from Carolina Biological Supply Co.
Anabaena	1/4S	Catalog #15-1225
Nostoc	1/4S	Catalog #15-1230
Oscillatoria	1/4S	Catalog #15-1235
Chlamydomonas	1/4S	Catalog #15-1245
Spirogyra	1/4S	Catalog #15-1320
Chlorella	1/4S	Catalog #15-1250
Ceratium	1/4S	Catalog #15-3245
Cyclotella	1/4S	Catalog #15-3020
Pond water	1/4S	Each group of students may be asked to bring their own sample to class.
Slides and coverslips	8/S	To be provided by student
Pasteur pipets	5/S	

PART IX
THE VIRUSES

EXERCISE 32
PHAGE ASSAY

Exercise 32 introduces the student to the plaque assay technique. We recommend that this exercise be performed one or two periods before other exercises in this section. The use of Escherichia coli B (ATCC 23226) and bacteriophage T2 (ATCC e11303-B2) has been very successful in our laboratory. It is recommended that this experiment be "tested" in the prep room before having the students do it to ascertain that the viral titer is sufficiently high and that the virus forms distinct plaques and not mottled ones. The student should be directed to the Appendix for the procedure for making dilutions and to exercise 7D to refamiliarize themselves with the proper pipeting technique.

OBJECTIVES:

A. Perform a plaque assay;

B. Familiarize the students with the appearance of viral "colonies" on bacterial lawns;

C. Illustrate that viruses are obligate intracellular parasites.

REQUIREMENTS FOR EACH LABORATORY SECTION

	QUANTITY	COMMENTS
Cultures		Obtain from ATCC or Carolina Biological Supply
Escherichia coli B	1/4S	ATCC 23226 CBS 12-4300
Bacteriophage T2	1/4S	ATCC e11303-B2 CBS 12-4325
or		
Bacteriophage T4	1/4S	ATCC 11303-B4 CBS 12-4330
Media		
Trypticase soy agar	3/2S	20 ml in 18x150 tubes
Soft (top) agar	3/2S	5 ml in 18x150 tubes
Dilution blanks	9/2S	9 ml saline in 16x100 tubes
Sterile 1ml pipets	9/2S	
Incubator	1/lab	Set at 35°C

EXERCISE 33
BACTERIOPHAGE ISOLATION

Exercise 33 introduces the student to a technique for the isolation of coliphages from nature. The students should be made aware that raw sewage and fresh manure may contain human pathogens and that care should be exercised when pipeting of handling them. If fresh manure is used, a slurry made by mixing equal parts (wt/vol) of manure and sterile water should be provided to the student in 100 ml screw-capped jars.

OBJECTIVES:

A.   Isolate a bacteriophage from sewage or manure;

B.   Learn techniques for the isolation of viruses from nature.

REQUIREMENTS FOR EACH LABORATORY SECTION

	QUANTITY	COMMENTS
Cultures Escherichia coli	1/4S	Obtain from ATCC Catalog #23226. 24-hour culture in TSB.
Raw sewage or fresh manure	1/2S	50 ml in screw-capped jars.

Media
  Trypticase soy agar
  Soft (top) agar
	4/S
	4/S
		20 ml in 18x150 tubes
		5 ml in 18x150 tubes

Erlenmeyer flask, 250 ml	1/2S	Sterile, with closure
Dilution blanks	9/2S	9 ml saline in 16x100 tubes
Sterile 1 ml pipets	9/2S	
Sterile 5 ml pipets	1/S	
Centrifuge	1/lab	
Centrifuge cups	1/2S	100 ml with closure, sterile
Incubator	1/lab	Set at 35° C

EXERCISE 34
PLANT VIRUS ISOLATION

Exercise 34 introduces the student to a technique for the isolation of the tobacco mosaic virus (TMV) from Nicotiana tabacum burley. The student will isolate the virus by preparing a slurry of diseased leaf tissue and then infect a variety of Nicotiana tabacum that show local lesions rather than mosiac disease. Either the xanthi or the glutinosa varieties will be suitable as hosts. The students can work individually or in pairs since one infected plant can be used by the entire class as the source of viruses.

OBJECTIVES:

A. Isolate the tobacco mosaic virus from tobacco plants;

B. Demonstrate the pathogenicity of the TMV on tobacco plants;

C. Show that different strains of plants respond differently to infection by the TMV.

REQUIREMENTS FOR EACH LABORATORY SECTION

	QUANTITY	COMMENTS
Tobacco plants		
Infected with TMV	1/class	_N. tabacum_ _burley_
Uninfected	1-2/class	_xanthi_ or _glutinosa_ varieties
Dilution blanks	5/2S	9 ml saline in 16x100 tubes
Sterile 1 ml pipets	5/2S	
Mortar and pestle	1/2S	

EXERCISE 35
ONE-STEP GROWTH CURVE

Exercise 35 is a modification of the classical experiment by Ellis and Delbruck demonstrating that if a population of cells is infected with a virus at a high multiplicity of infection so that each cell is almost certain to become infected at the same time, the life cycle of the virus will be synchronized and some important characteristics of the cycle can be observed. The student should be informed that lytic bacteriophages (e.g., T2) multiply fundamentally different from cells and as a result, the progeny from one virus can be several hundred rather than two, as is the case with cells. It is essential that the diluted samples be maintained in an ice bath after the incubation period to inhibit further the viral life cycle (step 4 of the first period). Also, the student (or group of students) must be thoroughly familiar with the experiment before they begin so that no time is lost "looking up" what to do next. The experiment must be performed in a prompt and efficient manner in order to obtain valid results.

OBJECTIVES:

A.   Determine the burst time of T2 bacteriophage;

B.   Determine the burst size of T2 bacteriophage;

C.   Perform the Ellis-Debruck experiment.

REQUIREMENTS FOR EACH LABORATORY SECTION

	QUANTITY	COMMENTS
Cultures Escherichia coli B	9.9 ml	ATCC 23226. A cell density of 5×10 /ml TSB is necessary. For initial infection.
Escherichia coli B	1/4S	Overnight cultures on TSB. For plaque assay.
Bacteriophage T2	0.1 ml	ATCC e11303-B2. A phage density of 5×10 /ml is suggested.
Media Trypticase soy broth	99 ml	In 250 ml flask with closure.
Trypticase soy agar	27/4S	20 ml in 18x150 tubes.
Soft (top) agar	27/4S	3 ml in 18x150 tubes.
KCI broth	9/4S	9 ml in 16x100 s/c tubes
KCI broth	27/4S	0.9 ml in 16x100 tubes
Petri plates	27/4S	
1 ml pipets	40/4S	
Water bath at 35°C	1/class	
Ice bath	1/4S	

EXERCISE 36
PHAGE TYPING OF
STAPHYLOCOCCUS AUREUS

Exercise 36 is a simplified version of the phage typing procedure for Staphylococcus aureus used in epidemiological studies. This exercise can be performed individually or in groups, one strain/group and the results then shared with the rest of the class. It uses three different Staphylococcus aureus phage types and four different typing phages. Exercise 36 can be scheduled in close proximity to other exercises in this part (IX) or with those in the medical microbiology section (XVI).

OBJECTIVE:

A. Determine the phage type of an unknown strain of Staphylococcus aureus.

REQUIREMENTS FOR EACH LABORATORY SECTION

	QUANTITY	COMMENTS
Cultures		
<u>Staphylococcus</u> <u>aureus</u>	1/4S	ATCC 27691
<u>Staphylococcus</u> <u>aureus</u>	1/4S	ATCC 27693
<u>Staphylococcus</u> <u>aureus</u>	1/4S	ATCC 27697
Bacteriophage type 47	1/4S	ATCC e27691-B1
Bacteriophage type 52A	1/4S	ATCC e27693-B1
Bacteriophage type 71	1/4S	ATCC e27697-B1
Bacteriophage type 81	1/4S	ATCC e27701-B1
Media		
Trypticase soy agar	1/S	20 ml in 18x150 tubes
Petri plates	1/S	
1 ml pipets	1/S	
Glass spreader	1/S	In alcohol jars
Sterile capillary pipets	4/S	

PART X
GENETICS OF MICROORGANISMS

EXERCISE 37
PHENOTYPIC AND GENOTYPIC VARIATION

Exercise 37 is a four-part exercise demonstrating the effect of the environment on the expression and mutation of bacterial genes. The first experiment demonstrates that the pigment produced by _Serratia marcescens_ (prodigiosin) is produced only at certain environmental temperatures (room temperature but not 35° C). The other three experiments are involved in the induction and selection of mutants. Each of these experiments include portions that demonstrate that unlike phenotypic variation, gene mutations are permanent changes.

OBJECTIVES:

A.   Demonstrate the phenomenon of phenotypic variation;

B.   Induce colorless mutations on genes that control bacterial pigmentation;

C.   Select for and isolate a streptomycin-resistant strain of _Escherichia coli_.

REQUIREMENTS FOR EACH LABORATORY SECTION

	QUANTITY	COMMENTS
Exercise 37A:		
Cultures		
Serratia marcescens	1/4S	Overnight cultures on TSB
Media		
Trypticase soy agar	6/S	20 ml in 18x150 tubes
Petri plates	6/S	
Incubator	1/lab	Set at 35°C
Room temperature incubator	1/lab	
Exercise 37B:		
Cultures		
Serratia marcescens	1/4S	Overnight cultures on TSB
Media		
Trypticase soy agar	6/S	20 ml in 18x150 tubes
Petri plates	6/S	
Incubators	2/lab	Set at 25°C and 35°C
Glass spreaders	1/S	In alcohol jars
Ultraviolet light box	1/4S	
Protective glasses	1/S	
Exercise 37C:		
Cultures		
Escherichia coli	1/4S	Overnight cultures on TSB
Media		
Trypticase soy agar	1/S	20 ml in 18x150 tubes
TSA+0.1 mg/ml streptomycin	1/S	20 ml in 18x150 tubes
Petri plates	1/S	

REQUIREMENTS FOR EACH LABORATORY SECTION

	QUANTITY	COMMENTS
Incubator	1/lab	Set at 35° C
Exercise 37D:		
Cultures Escherichia coli	1/S	

Media
  Trypticase soy agar
  TSA+0.1 mg/ml
    streptomycin
	2/S
	2/S
		20 ml in 18x150 tubes
		20 ml in 18x150 tubes

Petri plates	2/S	
Replica-plating blocks	1/S	Should fit snugly within bottom of Petri dish.
Velveteen squares	2/S	Sterile
Heavy rubber bands	1/S	
Incubator	1/lab	Set at 35°C

EXERCISE 38
GENE REGULATION

Exercise 38 introduces the student to the process of gene regulation by bacteria. The lactose operon is used because it is a well-studied model and the results are easy to obtain and usually interpretable. The experiment investigates the role of five different substances on the expression of the lactose operon. All should be performed because very important information about the functioning of the lactose operon is generated by the experiments. In order to economize in times and supplies each student (or group) should conduct the experiment using only one of the flasks (lactose, or glucose, or glucose + cAMP, or lactose + glucose, or TMG) and then share the results with the rest of the class. The control flask should be tested by every group to ascertain that their technique was adequate.

OBJECTIVES:

A.   Demonstrate the phenomenon of gene regulation;

B. Determine the effect of lactose, glucose, cAMP, and TMG on the expression of the lactose operon.

REQUIREMENTS FOR EACH LABORATORY SECTION

	QUANTITY	COMMENTS
Cultures Escherichia coli 　K12 lac+	5ml/S	ATCC e 23725. Overnight cultures on mineral-glycerol broth (MGB)*
Media* MGB+0.05% yeast 　extract	2/S	50 ml in 500 ml flasks
Reagents* 10% sucrose solution	2ml/S	Filter-sterilized
10% lactose solution	2ml/S	Filter-sterilized
10^{-1}M cAMP solution	2ml/S	Filter-sterilized
10% TMG	2ml/S	Filter-sterilized
2% potassium 　carbonate	5ml/S	
1% ONPG	2ml/S	
Toluene	1ml/S	
5 ml pipets	4/S	
Spectronic 20 　or	4/lab	Set at 600nm
Klett-Sommersen 　colorimeter	4/lab	
Parafilm	1 box/lab	Cut into squares
Water bath	1/lab	Set at 35°C

*See Appendix in lab manual for formulation.

EXERCISE 39
TRANSFORMATION

Exercise 39 introduces the student to the form of genetic recombination known as transformation. We use tryptophan auxotrophs (trpA23, trpB18, or trpE27) of <u>Acinetobacter</u> <u>calcoaceticus</u> because they are easy to work with and the results they provide are quite satisfactory and easily obtained. These mutants are available through the ATCC. The prototroph strain of these mutants is ATCC 33305. For further information on these organisms see: J. Bacteriol. 797-805 (1972). The exercise is sufficiently simple and it can be performed by each student in the class. It should be noted that the procedure (probably a modification of Griffith's work in the 1920's) does not involve the extraction and separation of DNA from the rest of the protoplasm. The student should be made aware of the classical studies by Avery, MacLeod, and McCarty, in which they demonstrated that DNA was the transforming principle elucidated by Griffith.

OBJECTIVES:

A. Demonstrate that competent bacteria can take up pieces of foreign DNA present in the bacterial environment;

B. Demonstrate that the uptake of naked DNA from a donor can sometimes be incorporated in the recipient's genome and become transformed.

REQUIREMENTS FOR EACH LABORATORY SECTION

	QUANTITY	COMMENTS
Cultures		
Acinetobacter calcoaceticus	1/2S	ATCC 33305. Wild type. Overnight cultures in TSB.
A. calcoaceticus trp	1/2S	ATCC 33306, 33307, or 33308. Tryptophan auxotrophs. Overnight cultures in TSB.
Media		
Mineral-Na acetate agar*	4/S	20 ml in 18x150 tubes
Lysis mixture*	1/S	Sterile, 25ml in 100ml flask
Sterile 16x150mm s/c tubes	4/S	
10 ml pipets	3/S	
1 ml pipets	4/S	
Capillary tubes	3/S	Sterile
Glass spreaders	2/S	In alcohol jars
Centrifuge	1/lab	
Centrifuge tubes, 10-25 ml	1/S	Sterile, with closure.
Water bath	1/lab	Set at 60°C
Ice water bath	4/lab	
Incubator	1/lab	Set at 35°C

*See Appendix for formulation.

EXERCISE 40
TRANSDUCTION

Exercise 40 introduces the student to the form of genetic recombination known as transduction. We use the Salmonella typhimurium-P22 bacteriophage model because it is extensively used and the results are usually very satisfactory. The prototroph and auxotroph mutants of S. typhimurium and P22 bacteriophage are readily available through ATCC. Any of the LT2-derived mutants (see ATCC catalog) can be used. The exercise should be performed in groups since it may be too complicated for a single student to perform.

OBJECTIVES:

A. Demonstrate that bacteria can take up pieces of foreign DNA carried inside a virus;

B. Demonstrate that the uptake of virus-carried DNA from a donor (wild type) can sometimes be incorporated in the recipient's genome and acquire a new phenotype.

EXERCISE 41
CONJUGATION

Exercise 41 introduces the student to the form of genetic recombination known as conjugation. The organism used is _Escherichia coli_ because there are numerous mutants (Hfr, F', and F⁻) available through ATCC. We introduce the student to the technique of gene mapping using both interrupted matings (B) and F' strains (A). We recommend that only one experiment (A or B) be carried out, in groups, during a given laboratory period and that no other new exercise be scheduled at that time.

OBJECTIVES:

A. Demonstrate that bacteria can transfer DNA from one cell to another;

B. Demonstrate that the uptake of bacterial DNA by an F⁻ strain from a donor (wild type) can sometimes be incorporated in the recipient's genome and acquire a new phenotype;

C. Show that the time required for transferring DNA can be used to locate the position of the gene in the donor's genome.

REQUIREMENTS FOR EACH LABORATORY SECTION

	QUANTITY	COMMENTS
Exercise 41A:		
Cultures		
Escherichia coli F' strains	1/4S	Overnight cultures in minimal-glucose broth*
F'lacY⁻thi⁻		ATCC e25251
F'lacZ⁻thi⁻		ATCC e25255
Escherichia coli F⁻ strains	1/4S	Overnight cultures in minimal-glucose broth*
F⁻lacZ⁻thi⁻		ATCC e25253
F⁻lacZ⁻thi⁻		ATCC e23722
F⁻lacY⁻thi⁻thr⁻leu⁻		ATCC e23724
F⁻lac⁻(deletion IZ)thi⁻		ATCC e23735
Media		
KCI broth*	4/S	Sterile, 100 ml in 250 ml flask
Minimal 0.4% lactose agar*	4/S	20 ml in 18x150 tubes. Supplemented with thiamine (= 0.1 µg/ml), threonine and leucine (each at 0.1 µg/ml).
Petri plates	4/S	
Sterile capillary tubes	4/S	
1 ml pipets	4/S	
Incubator	1/lab	Set at 35°C
Exercise 41B:		
Cultures		
Escherichia coli Hfr str^s	1/4S	CGSC 5461 Overnight culture in NB.
E. coli F⁻gal⁻lac⁻leu⁻str^r	1/4S	CGSC 5581 Overnight culture in TSB.

REQUIREMENTS FOR EACH LABORATORY SECTION

	QUANTITY	COMMENTS
Media		
Minimal broth*	6/4S	20 ml in 18x150 tubes
Minimal agar (MA) medium*	6/4S	20ml in 18x150 tubes. With thiamine.
MA+lac+str*	6/4S	20 ml in 18x150 tubes
MA+gal+str*	6/4S	20 ml in 18x150 tubes
MA+glu+str*	6/4S	20 ml in 18x150 tubes
Petri plates	24/4S	
Vortex mixer	4/lab	
1 ml pipets	6/4S	Sterile
1 ml pipets	2/4S	Sterile
Water bath	1/lab	
Incubator	1/lab	Set at 35°C
Colony counter	4/lab	

*See Appendix for formulation.

EXERCISE 42
AMES TEST

Exercise 42 introduces the student to the procedure known as the Ames test. This test is designed to determine the potential carcinogenic properties of substances. Each student should test a different carcinogen and then compare the results with those obtained by the rest of the class. The student should be warned that Salmonella typhimurium is a human pathogen and that care should be exercised when handling the cultures.

OBJECTIVE:

A. Demonstrate the carcinogenic potential of various substances by the Ames test.

REQUIREMENTS FOR EACH LABORATORY SECTION

	QUANTITY	COMMENTS
Cultures _Salmonella_ _typhimurium_	1/4S	(ATCC e29631) his Overnight culture in TSB.
Media Minimal glucose agar*	2/S	20 ml in 18x150 tubes.
Petri plates	2/S	
Biotin-histidine solution*	1/S	1 ml in 13x100 s/c tubes
1 ml pipets	2/S	
Propipet	1/S	

Mutagens
  hair dye
  old motor oil
  slurry of cigarette
    ashes
  slurry of chimney
    ashes
	1/S	1 ml in 13x100 s/c tubes
		
		1:1 (vol/wt) in water
		
		1:1 (vol/wt) in water

Sterile water		1 ml in 13x100 s/c tubes
Filter paper disks	1-4/S	
Forceps	1/S	
Sterile test tubes	1/S	
Water bath at 45°C	1/lab	
Water bath at 60°C	1/lab	
Incubator	1/lab	Set at 35°C

*See Appendix for formulation.

PART XI
MICROBIAL ECOLOGY

EXERCISE 43
POPULATION DYNAMICS (GROWTH CURVE)

Exercise 43 involves the construction of a bacterial growth curve. In this exercise, the population size will be measured by turbidimetric methods and viable plate counts. We suggest that students work in groups of four or six and that they develop a schedule so that the size of the population be measured at 30-60 minutes intervals for 6-12 hours. We suggest using a Klett-Summerson colorimeter and a side-arm flask because the results obtained using that system are quite reproducible. If side arm flasks are used (instead of cuvettes) instruct the students to "pour" back into the flask the contents of the side arm each time they make a measurement. If a spectronic 20 is used instead, refer the student to exercise 15 for instructions on how to use and make measurements with that colorimeter.

OBJECTIVES:

A. Determine the kinetics of a population of <u>Escherichia coli</u> growing in nutrient broth (NB);

B.   Compare the results obtained using O.D. and viable counts.

REQUIREMENTS FOR EACH LABORATORY SECTION

	QUANTITY	COMMENTS
Cultures		
Escherichia coli	1/4S	Overnight cultures in NB
Media		
Plate agar count	10/S	20 ml in 18x150 tubes
Nutrient broth	1/4S	25 ml NB+1% glucose in 250 ml side arm flasks
Petri plates	10/S	
Dilution blanks	10/S	9 ml saline in 16x150 tubes
1 ml pipets	10/S	
Shaking water bath	1/lab	Set at 37°C
Incubator	1/lab	Set at 35°-37°C

EXERCISE 44
MICROBIAL ANTAGONISM

Exercise 44 illustrates two aspects of microbial antagonism; competition and ammensalism. Both of these exercises can be performed by each student (or small groups) during the same laboratory period. The organisms suggested in the exercise are by no means the only ones suitable. We have included these because the results obtained using them are usually clear cut and reproducible.

OBJECTIVES:

A.   Illustrate microbial antagonism;

B.   Compare competition with ammensalism.

REQUIREMENTS FOR EACH LABORATORY SECTION

	QUANTITY	COMMENTS
Exercise 44A:		
Cultures		
Escherichia coli	1/4S	Overnight cultures in NB* + 1% glucose.
Staphylococcus aureus	1/4S	Overnight cultures in NB + 1% glucose.
Media		
NB + 1% glucose	1/S	25 ml in 250 ml flasks
Eosin Methylene blue agar	3/S	20 ml in 18x150 tubes
Mannitol-salts agar	3/S	20 ml in 18x150 tubes
Dilution blanks (.85% saline)	5/S	9 ml in 16x150 ml s/c tubes
Petri plates	6/S	
1 ml pipets	6/S	
Shaking water bath	1/lab	Set at 37°C
Incubator	1/lab	Set at 35°-37°C
Exercise 44B		
Cultures		
Escherichia coli	1/4S	Overnight cultures in NB* + 1% glucose.
Staphylococcus epidermidis	1/4S	Overnight cultures in NB + 1% glucose.
Pseudomonas fluorescens	1/4S	Overnight cultures in NB* + 1% glucose.
Bacillus cereus	1/4S	Overnight cultures in NB* + 1% glucose.
Penicillium sp.	1/4S	5-day cultures in Sab[#] broth+
Media		
Nutrient agar	9/S	28 ml in 18x150 tubes.

REQUIREMENTS FOR EACH LABORATORY SECTION

	QUANTITY	COMMENTS
Petri plates	9/S	
1 ml pipets	3/S	
Incubator	1/lab	Set at 30°C

*NB = nutrient broth (see Appendix for formulation).
#Sab = Sabouraud dextrose broth (see Appendix for formulation).

EXERCISE 45
MICROBIAL SYNERGISM

Exercise 45 illustrates the phenomenon of microbial synergism. This exercise can be performed by each student (or small groups), along with exercise 44, during a single laboratory period. Brom cresol purple (BCP) or phenol red (PR) can be used as pH indicators in the carbohydrate fermentation broths. Durham tubes consisting of 10x75 mm borosilicate glass tubes must be included in the culture medium.

OBJECTIVE:

A.   Illustrate microbial synergism.

REQUIREMENTS FOR EACH LABORATORY SECTION

	QUANTITY	COMMENTS
Cultures		
Escherichia coli	1/4S	Overnight cultures in NB*
Staphylococcus aureus	1/4S	Overnight cultures in NB
Proteus vulgaris	1/4S	Overnight cultures in NB
Media		
BCP-lactose broth	6/S	20 ml in 18x150 tubes, with Durham tubes.
BCP-sucrose broth	6/S	20 ml in 18x150 tubes, with Durham tubes.
Incubator	1/lab	Set at 37°C

*NB = nutrient broth (see Appendix for formulation).

EXERCISE 46
MICROBIAL POPULATIONS IN SOIL

Exercise 46 provides a means for the students to study some of the many microbial populations found in soils. The exercise will apply some of the selective techniques learned in other sections of the lab manual (e.g., exercise 7E) to isolate and characterize selected microbial groups from soils. The students can work in small groups and select for a particular organism, each group isolating a different one. At the completion of the exercise, each group can share and discuss their results with the rest of the class.

OBJECTIVES

A.   To isolate microbial populations from soils;

B.   Apply selective and differential cultural techniques for the isolation of microorganisms.

REQUIREMENTS FOR EACH LABORATORY SECTION

	QUANTITY	COMMENTS
Exercise 46A:		
Soil suspension	1/4S	11 g soil in 99 ml water
Media		
Plate count agar	4/4S	20 ml in 18x150 tubes
Sabouraud glucose agar	4/4S	20 ml in 18x150 tubes
Starch-Casein agar	4/4S	20 ml in 18x150 tubes
Petri dishes	12/4S	
Dilution blanks (saline)	7/4S	9 ml in 16x150 s/c tubes.
Cyclohexamide stock solution	1/4S	5 ml in 16x150 s/c tubes.
Pipets, 1ml	8/4S	
Incubator	1/lab	Set at 30°C
Exercise 46B:		
Soil sample	1/4S	See above (46A)
Media		
Plate count agar	4/4S	20 ml in 18x150 tubes
Sporulation agar	4/4S	20 ml in 18x150 tubes
Reagents		Available in lab bench
Gram stain		
Endospore stain		
Dilution blanks (saline)		See above (46A)
Incubator	1/lab	Set at 30°C
Exercise 46C:		
Cultures		
Streptomyces sp	1/4S	Isolated in 46A above.
Escherichia coli	1/4S	Overnight cultures in TSB.
Staphylococcus aureus	1/4S	Overnight cultures in TSB.
Pseudomonas fluorescens	1/4S	Overnight cultures in TSB.

REQUIREMENTS FOR EACH LABORATORY SECTION

	QUANTITY	COMMENTS
Media		
Plate count agar	1/4S	20 ml in 18x150 tubes.
Incubator	1/lab	Set at 30°C

EXERCISE 47
THE NITROGEN CYCLE

Exercise 47 introduces the student to some of the biological conversions of the element nitrogen in the nitrogen cycle. Four aspects of the nitrogen cycle will be examined: ammonification, nitrification, denitrification, and nitrogen fixation. The exercises, as described in the manual, are relatively simple to perform and can be done in one laboratory period. We suggest that each student (or small group) perform the exercise and then discuss the results with the rest of the class. Exercise 47B involves the use of a highly concentrated (1:3 dilution) sulfuric acid as a reagent. It might be safer if the students would bring the spot plates to the instructor's desk and ask the instructor to dispense the sulfuric acid instead of handling this very corrosive reagent on their own.

OBJECTIVES:

A. To illustrate how microorganisms recycle nutrients in ecosystems;

B.   Study some of the various components of the nitrogen cycle.

REQUIREMENTS FOR EACH LABORATORY SECTION

	QUANTITY	COMMENTS
Exercise 47A:		
Cultures		
Escherichia coli	1/4S	Overnight cultures in TSB.
Pseudomonas fluorescens	1/4S	Overnight cultures in TSB.
Garden soil	1/4S	1 g in a zip-lok plastic bag
Media		
4% peptone broth*	3/S	20 ml in 18x150 tubes.
Reagents		
Nessler's*	1/4S	In brown bottles with dropper.
Pasteur pipets with bulbs	3/S	
Spot plates	1/4S	
Incubator	1/lab	Set at 30°C
Exercise 47B:		
Garden soil	1/4S	1 g in a zip-lok plastic bag
Media		
Ammonium broth*	1/S	20 ml in 6 oz prescription bottles.
Nitrite broth*	1/S	20 ml in 6 oz prescription bottles.
Reagent		
Nessler's	1/4S	In brown bottles with dropper
Trommsdorf's*	1/4S	In brown bottles with dropper
Diphenylamine	1/4S	In brown bottles with dropper
Sulfuric acid (1:3)	1/4S	In spill-proof containers with suitable dispensing pipets.
Pasteur pipets with bulbs	3/S	

REQUIREMENTS FOR EACH LABORATORY SECTION

	QUANTITY	COMMENTS
Spot plates	1/4S	
Incubator	1/lab	Set at 18°-20°C
Exercise 47C:		
Cultures		
Escherichia coli	1/4S	Overnight cultures in TSB.
Pseudomonas denitrificans	1/4S	Overnight cultures in TSB.
Garden soil	1/4S	1 g in zip-lok plastic bag
Media		
Nitrate broth*	3/S	20 ml in 18x150 tubes with Durham tubes.
Reagents		
Nessler's	1/4S	In brown bottles with dropper
Sulfamic acid*	1/4S	In brown bottles with dropper
Zinc dust	1/4S	In small vials with spatula
Pasteur pipets with bulbs	3/S	
Spot plates	1/4S	
Incubator	1/lab	Set at 30°C
Exercise 47D:		
Legume with root nodules	1/4S	
Garden soil	1/4S	5g in a zip-lok plastic bag
Media		
Nitrogen-free broth*	1/S	50 ml in 250 ml flasks
Reagents		
Gram staining	1/4S	Available in lab bench
Forceps	1/S	
Glass slides	4/S	Student should provide own
Pasteur pipets with bulbs	1/S	

REQUIREMENTS FOR EACH LABORATORY SECTION

	QUANTITY	COMMENTS
Spot plates	1/4S	
Incubator	1/lab	Set at 30°C

*See Appendix for formulation.

EXERCISE 48
THE SULFUR CYCLE

Exercise 48 introduces the student to some of the biological conversions of the element sulfur in the sulfur cycle. The first part of the exercise involves the construction of a Winogradsky column. It will serve as a good source of microorganisms for subsequent parts of this exercise. We suggest that the column be started at least three weeks prior to the performance of the other parts of this exercise. The other three parts: examination of photosynthetic bacteria, isolation of <u>Thiobcillus</u>, and isolation of sulfate reducers, can be started during the same laboratory period. Small groups of students (2-4) can perform the exercise and then share and discuss their results at the conclusion of the exercise.

OBJECTIVES:

A. To illustrate how microorganisms recycle nutrients in ecosystems;

B. Study some of the various components of the sulfur cycle.

REQUIREMENTS FOR EACH LABORATORY SECTION

	QUANTITY	COMMENTS
Exercise 48A:		
Bottom ooze (mud)	40g/4S	In plastic bag or bucket
Decomposed plant material	10g/4S	In plastic bag or bucket
Pond water	50ml/4S	In screw-capped jar
1000 ml graduated cylinder	1/4S	
Calcium sulfate	50g/4S	
Plastic wrap	1 roll/lab	
Rubber bands	1/4S	
Aluminum foil	1 roll/lab	
Incandescent light	4/lab	With 60 watt bulb.
Plastic bucket	1/4S	To mix the column's contents
Exercise 48B:		
Modified van Niel's medium*	1/4S	In B.O.D. bottles with ground glass stoppers.
Winogradsky column	1/4S	From exercise 48A. Or pond water.
Pasteur pipets, curved	1/S	Flame tip to curve them.
Depression slides	1/S	To be provided by student.
Coverslips	2/S	To be provided by student.
Glass slides	2/S	To be provided by student.
Gram staining reagents	1/4S	Available in lab.
Incandescent light source	1/4S	

REQUIREMENTS FOR EACH LABORATORY SECTION

	QUANTITY	COMMENTS
Exercise 48C:		
Winogradsky column	1/4S	From exercise 48A. Or soil or pond water.
Media		
Modified Starkey's*	1/S	4 ml in 13x100 s/c tubes
Thiosulfate medium, pH 7.0*	1/S	25 ml in 100 ml flask
Thiosulfate medium, pH 4.5*	1/S	25 ml in 100 ml flask
Exercise 48D:		
Winogradsky column	1/4S	From exercise 48A. Or soil or pond water.
0.5% ferrous ammonium sulfate	1/S	1 ml in 13x100 s/c tubes
Desulfovibrio medium*	1/S	10 ml in 16x150 s/c tubes
Dilution blanks, saline	6/4S	9 ml in 16x150 s/c tubes
1 ml pipets	6/4S	

*See Appendix for formulation.

EXERCISE 49
LUMINOUS BACTERIA

Exercise 49 involves the isolation of luminescent bacteria from marine fishes. Bottom-dwelling fishes are a good source of the bioluminescent bacteria. The fish (e.g., cod or red snapper) should be "incubated" in a cold room for two days to a week and examined in the dark for luminescent bacteria. The instructor can do this ahead of time so that the students can proceed with the isolation of the bacteria. Each student should isolate his/her own bacterium because not all isolation attempts will be successful. About 50% of the class, however, should isolate a luminous bacterium. The results can be shared and discussed at the end of the exercise.

OBJECTIVES:

A.   Isolate bioluminescent bacteria from bottom-dwelling fish;

B.   Illustrate that cold incubation can select for psychrophilic bacteria.

REQUIREMENTS FOR EACH LABORATORY SECTION

	QUANTITY	COMMENTS
Dead marine fish	1/lab	
Salt water	100 ml	
Sterile 3% NaCl	1/S	5 ml in 13x100 s/c tubes
Peptone-glycerol phosphate-NaCl (PGPS) agar*	1/S	20 ml in 18x150 tubes
PGPS broth*	1/S	20 ml in 18x150 tubes
Refrigerator	1/lab	Set at about 10°C

*See Appendix for formulation

EXERCISE 50
PERIPHYTIC BACTERIA

Exercise 50 involves the isolation of periphitic bacteria from aquatic environments using glass slides. Each student can prepare his or her own slide, labelling it with a diamond-tipped pencil, and inserting it in a fish tank containing river or pond water. The slides can be observed after 7-10 days and the results shared and discussed at that time.

OBJECTIVES:

A.   Isolate periphitic bacteria from river or pond water;

B.   Illustrate the bacterial diversity in aquatic environments.

REQUIREMENTS FOR EACH LABORATORY SECTION

	QUANTITY	COMMENTS
Fish tank with pond water	1/lab	Use river or pond water. Insert a test tube rack in the tank to place the slides. More than one tank, each with a different source of water may be used.
Microscope slides	1/S	To be provided by student.
Gentian violet	1/4S	1:50 dilution using tap water as diluent. In brown bottles with dropper.

PART XII
WATER SANITARY ANALYSIS

EXERCISE 51
MOST PROBABLE NUMBER METHOD
FOR COLIFORM ANALYSIS

Exercise 51 is the multiple tube fermentation test for coliforms as outlined in Standard Methods for the Analysis of Water and Wastewater. It should be pointed out that in this procedure, the confirmed test involves the inoculation of brilliant green bile lactose broth (with Durham tubes) using as inoculum positive presumptive tubes. Students may be asked to bring their own sample of drinking water (a minimum of 50 ml) for testing.

OBJECTIVE:

A. Determine the coliform MPN of a drinking water sample.

REQUIREMENTS FOR EACH LABORATORY SECTION

	QUANTITY	COMMENTS
Cultures		
Escherichia coli	1/4S	Overnight cultures on TSA. Use as control for the various culture media used.
Water sample	1/4S	50-100 ml in 125 ml wide mouth jars.
Media		
2x Lauryl tryptose broth*	6/S	10 ml in 20x150 tubes with Durham tubes.
E.C. broth*	6/S	10 ml in 20x150 tubes with Durham tubes.
#BGBL broth*	6/S	10 ml in 20x150 tubes with Durham tubes.
#EMB agar	2/S	20 ml in 18x150 tubes
Nutrient agar slants	2/S	20 ml in 18x150 tubes with a 1" butt.
Gram staining reagents	1/4S	Available on lab bench
Water bath	1/lab	Set at 44.5°C
Incubator	1/lab	Set at 35°C

*See Appendix for formulation.
#BGBL = Brilliant Green Bile Lactose
 EMB = Eosin Methylene Blue

EXERCISE 52
MEMBRANE FILTER METHOD
FOR COLIFORM ANALYSIS

Exercise 52 is the membrane filter method for coliform analysis in potable water as outlined in Standard Methods for the Analysis of Water and Wastewater. Students may be asked to bring and analyze their own sample of drinking water (a minimum of 100 ml).

OBJECTIVE:

A. Determine the number of coliforms/100 ml and/or fecal coliforms/100 ml of water by the membrane filter method.

REQUIREMENTS FOR EACH LABORATORY SECTION

	QUANTITY	COMMENTS
Water sample	1/S	100 ml in 125 ml wide mouth jars of plastic bags. Should contain 10-50 coliforms/100 ml
Media m-Endo broth*	1/S	2.0 ml in 13x100 tubes. For totoal coliforms.
m-FC broth*	1/S	2.0 ml in 13x100 tubes. For fecal coliforms.
Membrane filter apparatus	1/4S	
Membrane filters (0.45 m)	1/S	
Forceps in alcohol jars	1/S	
Cultures dishes, 60x15 mm	1/S	
Absorbent pads, 48mm	1/S	
Incubator	1/lab	Set at 35° C

*See Appendix for formulation.

EXERCISE 53
NONCOLIFORM INDICATORS OF RECREATIONAL WATER QUALITY

Exercise 53 is a two-part exercise involving the detection of pseudomonads by the MPN method and staphylococci by the membrane filter method. These organisms are important indicators of recreational water pollution. This exercise may be performed in pairs, one student testing for pseudomonads and the other for staphylococci. Each student might be asked to bring their own water sample from a swimming pool, hot tub, or spa.

OBJECTIVES:

A. Determine the MPN/100 ml pseudomonads in recreational water;

B. Determine the number of staphylococci/100 ml of recreational water by the membrane filter method.

REQUIREMENTS FOR EACH LABORATORY SECTION

	QUANTITY	COMMENTS
Water sample (pool or spa)	1/S	100 ml in 125 ml wide mouth jars or plastic bags. Should contain 10-50 coliforms/100 ml
Cultures <u>Pseudomonas aeruginosa</u>	1/4S	Overnight cultures in TSA.

Media
  m-staphylococci
    broth*
	1/S	2.0 ml in 13x100 tubes. For total coliforms.

2x asparagine broth*	6/S	10 ml in 18x100 tubes
1x asparagine broth*	6/S	10 ml in 18x100 tubes
Acetamide broth*	6/S	10 ml in 18x100 tubes
Membrane filter apparatus	1/4S	
Membrane filters (0.45 m)	1/S	
Forceps	1/S	In alcohol jars
Culture dishes	1/S	60x15 mm
Absorbent pads	1/S	48 mm
Black light	1/lab	
Gram staining reagents	1/4S	Available on lab bench
3% hydrogen peroxide	1/4S	In brown bottles with dropper
Incubator	1/lab	Set at 35°C

*See Appendix for formulation

EXERCISE 54
ISOLATION OF *ACTINOMYCETES* FROM WATER AND WASTEWATER TREATMENT PLANTS

Exercise 54 involves the isolation of actinomycetes from water supplies, using cyclohexamide as a selective agent and starch-casein agar as a differential medium. The instructor may provide a sample of water from an activated sludge plant to each group of students or the students may prepare a slurry of soil (1:2) as the source of actinomycetes. The student should be warned that sewage generally contain pathogenic microorganisms and/or viruses, and therefore, must exercise care when handling the samples.

OBJECTIVE:

A.   Isolate actinomycetes from soil or wastewater.

REQUIREMENTS FOR EACH LABORATORY SECTION

	QUANTITY	COMMENTS
Water sample, activated sludge	1/S	10 ml in 50 ml wide mouth jars.
Media Starch-casein agar*	3/S	20 ml in 18x150 tubes
Petri dishes	3/S	
Cyclohexamide stock solution*	1ml/S	
Dilution blanks, 99 ml	1/S	
Dilution blanks, 9 ml	4/S	
Incubator	1/lab	Set at 28°C

*See Appendix for formulation

PART XIII
FOOD MICROBIOLOGY

EXERCISE 55
FOOD SPOILAGE

Exercise 55 applies the techniques for viable counts learned in exercise 14 and those earned in exercise 7E for selective and differential media to determine the presence of spoilage organisms in certain types of foods. Each student might be asked to bring his/her own sample of food and analyze it. Alternatively, the instructor may provide the class with a blended food sample diluted 1:10. Parts A, B, and C can be performed using the same food sample, although exercise 55C will work better if butter is used instead of hamburger meat. All four parts can be performed in a single laboratory period.

OBJECTIVES:

A. Enumerate food spoilage organisms in hamburger, butter, and fruits;

B. Apply selective techniques for the detection of specific microbial groups.

REQUIREMENTS FOR EACH LABORATORY SECTION

	QUANTITY	COMMENTS
Exercise 55A:		
Hamburger sample (1:10)	1/4S	Blend 50 g in 450 ml water
Plate count agar	3/S	20 ml in 18x150 tubes
99 ml dilution blanks	1/S	
9 ml dilution blanks	1/S	
1 ml pipets	4/S	
Petri plates	3/S	
Incubator	1/lab	Set at 35°C
Exercise 55B:		
Cultures Escherichia coli (-)	1/4S	Use as controls Overnight cultures on TSA
Bacillus subtilis (+)	1/4S	Overnight cultures on TSA
Plate count agar	4/S	20 ml in 18x150 tubes
Sterile skim milk	4/S	2 ml in 18x150 tubes
Incubator	1/lab	Set at 35°C
Exercise 55C:		
Cultures Escherichia coli	1/4S	Use as controls Overnight cultures on TSA
Pseudomonas fluorescens	1/4S	Overnight cultures on TSA
Fat agar plates*	5/S	20 ml in 18x150 tubes. Use medium used in exercise 23.
Petri plates	5/S	
Glass spreaders	4/S	In alcohol jars
Melted butter	1/S	Pads
99 ml dilution blanks	1/S	Keep warm (45°C)

REQUIREMENTS FOR EACH LABORATORY SECTION

	QUANTITY	COMMENTS
9 ml dilution blanks	1/S	Keep warm (45°C)
1 ml pipets	4/S	Keep warm (45°C)
Incubator	1/lab	Set at 35°C
Exercise 55D:		
Fruit sample (1:10)	1/4S	Blend 50 g in 450 ml water
Sabouraud dextrose agar	3/S	20 ml in 18x150 tubes
Petri plates	3/S	
1 ml pipets, sterile	3/S	
99 ml dilution blanks	1/S	
9 ml dilution blanks	3/S	
Glass spreaders	3/S	
Incubator	1/lab	Set at 25°C

*See Appendix for formulation

EXERCISE 56
FOOD POISONINGS AND INTOXICATIONS

Exercise 56 applies the techniques for viable counts learned in exercise 14 and those learned in exercise 7E for selective and differential media to determine the presence of staphylococci in certain types of foods. Each student might be asked to bring his/her own sample of food and analyze it. Alternatively, the instructor will provide the class with a blended food sample, diluted 1:10, and seeded with 1 ml of a 1/10,000 dilution of an overnight culture of S. aureus in TSB.

OBJECTIVES:

A. Enumerate staphylococci in foods;

B. Apply selective techniques for the detection of specific microbial groups.

REQUIREMENTS FOR EACH LABORATORY SECTION

	QUANTITY	COMMENTS
Cultures Staphylococcus aureus Staphylococcus epidermidis Escherichia coli	1/4S	Overnight cultures in TSB. Use as controls.
Food sample (1:10)	1/4S	Blend 50 g in 450 ml water. Seed with S. aureus
Mannitol-salts agar*	4/S	20 ml in 18x150 tubes
99 ml dilution blanks	1/S	
9 ml dilution blanks	4/S	
1 ml pipets with bulbs	4/S	
Petri plates	3/S	
Incubator	1/lab	Set at 35°C

*See Appendix for formulation

EXERCISE 57
EVALUATING THE
SANITARY QUALITY OF FOODS

Exercise 57 applies the techniques for viable counts learned in exercise 14 and those learned in exercise 7E for selective and differential media to determine the presence of coliforms and fecal streptococci in food samples. Each student might be asked to bring his/her own sample of food and analyze it. Alternatively the instructor will provide the class with a blended food sample, diluted 1:10, and seeded with 1 ml of a 1/10,000 dilution of an overnight culture of S. coli in TSB.

OBJECTIVES:

A. Enumerate coliforms and fecal streptococci in foods;

B. Apply selective techniques for the detection of specific microbial groups.

REQUIREMENTS FOR EACH LABORATORY SECTION

	QUANTITY	COMMENTS
Exercise 57A:		
Cultures	1/4S	Overnight cultures in TSB. Use as control.
Escherichia coli		
Food sample (1:10)	1/4S	Blend 50 g in 450 ml water. Seed with _E. coli_.
Violet red bile agar*	4/S	20 ml in 18x150 tubes
9 ml dilution blanks	4/S	
1 ml pipets with bulbs	4/S	
Petri plates	4/S	
Colony counter	4/lab	
Incubator	1/lab	Set at 35°C
Exercise 57B:		
Cultures	1/4S	Overnight cultures on BHI. Use as control.
Streptococcus faecalis		
Food sample (1:10)	1/4S	Blend 50 g in 450 ml water. Seed with _S. faecalis_.
Media		
KF streptococcal agar*	4/S	20 ml in 18x150 tubes
Bile esculin agar*	4/S	20 ml in 18x150 tubes
Brain-heart infusion	4/S	10 ml in 18x150 tubes
BHI+6.5% NaCl broth	4/S	10 ml in 18x150 tubes
9 ml dilution blanks (saline)	3/S	In 16x150 s/c tubes
1 ml pipets with bulbs	4/S	
Petri plates	4/S	
Colony counter	4/lab	

REQUIREMENTS FOR EACH LABORATORY SECTION

	QUANTITY	COMMENTS
3% hydrogen peroxide	4/lab	In brown bottle with dropper
Incubators	2/lab	Set at 35° and 45°C

*See Appendix for formulation

EXERCISE 58
EXAMINATION OF
EATING UTENSILS AND EQUIPMENT

Exercise 58 illustrates an accepted method for the evaluation of eating utensils in restaurants and cafeterias. The instructor may provide each student with a sterile cotton swab and a screw-capped tube (18x100 mm) containing 9ml of sterile 0.85% saline. The student may then sample eating utensils at home according to the instructions in the manual and complete the exercise on the following day or laboratory period.

OBJECTIVE:

A.   Establish the sanitary quality of eating utensils.

REQUIREMENTS FOR EACH LABORATORY SECTION

	QUANTITY	COMMENTS
Eating utensil	1/S	
Plate count agar*	3/S	20 ml in 18x150 tubes
9 ml dilution blanks	1/S	
1 ml pipets with bulbs	2/S	
Petri plates	2/S	
Rodac plate	1/S	
Colony counter	4/lab	
Incubator	1/lab	Set at 35°C

*See Appendix for formulation

EXERCISE 59
MICROBES AND FOOD PRODUCTION

Exercise 59 illustrates methods for the laboratory preparation and/or examination of sauerkraut, vinegar, and sourdough bread. The exercise illustrates the role of microorganisms in the production of foods, using these foods as examples. Each group of students can work in the preparation and/or analysis of the various foods and share the results with the rest of the class.

OBJECTIVES:

A. Prepare foods in which microbial participation is important;

B. Examine the various microbial groups that are involved in the production of sauerkraut, vinegar, and sourdough bread.

REQUIREMENTS FOR EACH LABORATORY SECTION

	QUANTITY	COMMENTS
Exercise 59A:		
Cabbage heads	2/4S	
Table salt	1 pound	
Sharp knife	1/4S	For shredding the cabbage.
Balance	4/lab	
Widemouth jars, 1 gallon	1/4S	
Plastic bags, 10"x24"	1/S	
Media Tomato juice agar	3/S	Make available for 4 periods
Dilution blanks	3/S	Make available for 4 periods
3% hydrogen peroxide	1/4S	In brown bottle with dropper
Phenol red indicator*	10ml/4S	
0.1N NaOH	10ml/4S	
5 ml pipets	1/4S	
Titration apparatus	1/4S	May use pH paper instead to determine that there is a decline in pH of brine as the population of lactic acid bacteria develops.
125 ml beakers	1/4S	
Incubator	1/lab	Set at 30°C
Exercise 59B:		
Apple juice	1/2S	30 ml in 25x150 tubes
Apple or raisins	1/2S	
Cultures Saccharomyces cerevisiae	1/4S	3-day cultures in Sab broth

REQUIREMENTS FOR EACH LABORATORY SECTION

	QUANTITY	COMMENTS
Acetobacter aceti	1/4S	3-day cultures in Sab broth
Unpasteurized vinegar	10 ml	
Exercise 59C:		
Sourdough starter	1/4S	
Media Sabouraud dextrose agar	1/S	20 ml in 18x150 tubes
Ethanol-calcium carbonate*	1/S	20 ml in 18x150 tubes
Gram staining agents	1/4S	Available in lab bench
Incubator	1/lab	Set at 30°C

*See Appendix for formulation

PART XIV
DAIRY MICROBIOLOGY

EXERCISE 60
ANALYSIS OF THE SANITARY QUALITY
OF MILK AND DAIRY PRODUCTS

Exercise 60 includes some of the methods used for the evaluation of the sanitary quality of milk as published by the American Public Health Association. Each of the tests outline in the manual are simple to do and can be done, by each student, in a single laboratory period. Commercially-available raw milk will contain sufficient organisms to provide suitable results. Since samples of milk may contain either too few or too many bacteria/ml, we suggest that a "test" run be carried out in the prep room before the exercise is carried out by the students.

OBJECTIVES:

A. Evaluate the sanitary quality of raw milk;

B. Become familiar with some of the procedures utilized in industry to evaluate dairy products for human consumption.

REQUIREMENTS FOR EACH LABORATORY SECTION

	QUANTITY	COMMENTS
Exercise 60A:		
Raw milk	1/4S	25 ml in wide mouth jars

Media
  Violet red bile
    agar*
	5/S	20 ml in 18x150 tubes. For best results, have students overlay about 10ml of VRBA over the hardened VRBA-milk plates.

Petri dishes	3/S	
9 ml dilution blanks	3/S	In 16x150 s/c tubes
1 ml pipets	3/S	
Incubator	1/lab	Set at 30°C
Exercise 60B:		
Raw milk	1/4S	25 ml in wide mouth jars
Standard resazurin solution*	1/4S	In brown bottle with dropper
16x150mm s/c tubes	2/S	Clean but not sterile
1 ml pipets	2/S	
Thermometer	1/4S	
Fluorescent light	1/4S	
Water bath	1/lab	Set at 36°C
Exercise 60C:		
Raw milk	1/4S	25 ml in wide mouth jars
Milk loops (0.02 ml)	1/S	Or 10 l Eppendorf pipet
Warming table	1/lab	Set at 40° to 45°C
Levowitz-Weber stain	1/4S	In brown bottle with stopper
500ml beaker	4/4S	With warm tap water. May use Caplin jars instead.

*See Appendix for formulation

EXERCISE 61
MASTITIS MILK

Exercise 61 includes some of the methods used for the detection of mastitis in dairy cattle as published by the American Public Health Association. Each of the tests outlined in the manual are simple to do and can be done, by each student, in a single laboratory period. Samples of milk from normal cows and cows with mastitis may be obtained from regional dairy quality monitoring agencies. As the CMT and the Whiteside test are indirect methods for the detection of mastitis, students may compare the results on the same sample of milk tested by these two methods as well as by the DMSCC.

OBJECTIVES:

A. Evaluate the sanitary quality of raw milk;

B. Become familiar with some of the procedures utilized in industry to detect mastitis in dairy cows.

REQUIREMENTS FOR EACH LABORATORY SECTION

	QUANTITY	COMMENTS
Exercise 61A:		
Raw milk	1/4S	25 ml in wide mouth jars. From normal and cows with various degrees of mastitis.
CMT reagent	1/4S	In small bottles.
Milk paddles	1/4S	
Exercise 61B:		
Raw milk	1/4S	25 ml in wide mouth jars. From normal and cows with various degrees of mastitis.
4% NaOH solution	1/4S	
Applicator stick	1/S	
Test plates	1/4S	
Exercise 61C:		
Raw milk	1/4S	25 ml in wide mouth jars. From normal and cows with various degrees of mastitis.
Reagents	1/4S	See exercise 60C.
Exercise 61D:		
Raw milk	1/4S	25 ml in wide mouth jars. From normal and cows with various degrees of mastitis.
Cultures		
Escherichia coli	1/4S	Overnight cultures in TSB
Staphylococcus aureus	1/4S	Overnight cultures in TSB
Pseudomonas aeruginosa	1/4S	Overnight cultures in TSB
Streptococcus agalactiae	1/4S	Overnight cultures in TSB
Media		
EMB agar*	2/S	20 ml in 18x150 tubes

REQUIREMENTS FOR EACH LABORATORY SECTION

	QUANTITY	COMMENTS
Mannitol salts agar*	2/S	20 ml in 18x150 tubes
Blood agar (5% SRBC)*	2/S	Poured in plates
Incubator	1/lab	Set at 35°C

*See Appendix for formulation

EXERCISE 62
PASTEURIZATION OF MILK

Exercise 62 involves the laboratory pasteurization test as performed by laboratories to evaluate the quality of raw milk. The timing of the pasteurization process should begin when the milk temperature reaches 62 degrees. It is advisable that a thermometer be inserted in a control, 10 ml sample of milk, along with test samples, to monitor the milk temperature. The exercise can be performed in small groups. An interpretation of the results should follow the conclusion of the exercise.

OBJECTIVE:

A.   Carry out the laboratory pasteurization test.

REQUIREMENTS FOR EACH LABORATORY SECTION

	QUANTITY	COMMENTS
Raw milk	1/S	10 ml in 18x125 s/c tubes
Thermometer	1/lab	Inserted in a tube with milk
Media		
Violet red bile agar*	3/S	20 ml in 18x150 tubes. Overlays of VRBA will improve the results.
Plate count agar*	3/S	20 ml in 18x150 tubes
Petri plates	5/S	
Dilution blanks	2/S	9 ml saline in 16x150 tubes
1 ml pipets	3/S	
Water bath	1/lab	Set at 63.5°C
Incubator	1/lab	Set at 32°C

*See Appendix for formulation

EXERCISE 63
MICROBES AND THE
PRODUCTION OF DAIRY FOODS

Exercise 63 involves the preparation and examination of dairy foods such as cheese and yogurt. A single batch of each can be prepared in the laboratory and the results shared with the entire lab section. This exercise can be scheduled during the same laboratory period in which exercise 59 is scheduled.

OBJECTIVES:

A.   Prepare yogurt and cheese in the laboratory;

B.   Evaluate the role of the lactic acid bacteria in the production of dairy foods.

REQUIREMENTS FOR EACH LABORATORY SECTION

	QUANTITY	COMMENTS
Exercise 63A:		
Pasteurized milk	1 quart	
Powdered milk	1 box	To make 1 quart
Plain yogurt	1 pint	
Balance	1/lab	
Gram staining reagents	1/4S	Available in lab bench.
Cups	1/S	Plastic or styrofoam
Aluminum foil	1 roll	
Magnetic stirrer and bar	1/lab	
Incubator	1/lab	Set at 46°C. An oven with a gas pilot will also be adequate.
Exercise 63B:		
Pasteurized milk	1 quart	
Buttermilk	1 quart	
Cheeses	1 each	Various types of cheeses
Cheese cloth	1 roll	
Media Gelatin-peptone agar*	1/S	20 ml in 18x150 tubes
Acidified potato dextrose*	1/S	20 ml in 18x150 tubes
Gram staining reagents	1/4S	
Incubators	2/lab	Set at 30° and 37°C

*See Appendix for formulation

PART XV
AGRICULTURAL MICROBIOLOGY

EXERCISE 64
PLANT PATHOGENS

Exercise 64 introduces the student to some plant diseases caused by bacteria and fungi. Exercise 64C, crown gall of tomato, is an excellent experiment for illustrating Koch's postulates. If only one experiment in this exercise can be performed, exercise 64C should be the one. Exercise 64B will produce malodorous results and we suggest that the potatoes be kept outside the laboratory so that the smell will not cause discomfort to the students (and instructor). The pathogen suspensions in exercises 64C and 64D may be prepared by the instructor during the lab session (as a demonstration) and then 1 ml aliquots distributed to the students in sterile vials.

OBJECTIVES:

A. Examine diseased plants and the pathogens;

B. Demonstrate Koch's postulates.

169

REQUIREMENTS FOR EACH LABORATORY SECTION

	QUANTITY	COMMENTS
Exercise 64A:		

Prepared slides of
  Puccinia
  Ustilago
	1/4S	Obtain from Carolina Biological Supply Co.

Specimens
  Infected barberry
  Infected wheat
  Infected corn
	1/4S	

Exercise 64B:		
Small, firm potatoes	1/S	
Carrot rounds	1/S	

Cultures
  Erwinia caratovora
  Erwinia atroseptica
	1/4S
	1/4S
		Overnight cultures in TSB
		Overnight cultures in TSB

Petri dishes	2/S	
Sterile distilled water	1/S	10 ml in 18x150 tubes
Scalpel	1/4S	
Forceps	1/4S	
Pasteur pipets	1/S	
Filter paper	1/S	
Exercise 64C:		

Tomato plants
  Healthy
  With crown gall
	4/lab	

Trypticase soy agar	2/S	20 ml in 18x150 tubes
Petri plates	2/S	
Mortar and pestle	1/4S	
Sterile saline	1/S	10 ml in 18x150 tubes
Alcohol jars	1/4S	

REQUIREMENTS FOR EACH LABORATORY SECTION

	QUANTITY	COMMENTS
Scalpel	1/4S	
Dissecting needles	1/4S	
Gram staining reagents	1/4S	
Exercise 64D:		

Tobacco plants
  Healthy
  With mosaic disease
	4/lab	

Blender	1/lab	
Sterile saline	1/S	10 ml in 18x150 tubes
Funnel	1/4S	
Filter paper	1/4S	
Membrane filter apparatus	1/4S	With 0.45μm membrane filter
Carborundum	1 jar	
Cotton (or Dacron) swabs	1/S	

*See Appendix for formulation

EXERCISE 65
MYCORRHIZAE

Exercise 65 introduces the student to the symbiotic association between fungi and plants known as mycorrhiza. We recommend the use of pine seedling root hairs as an excellent source of mycorrhizae. Endothrophic mycorrrhizae are harder to find and demonstration slides purchased from Caroline Biological or other biological supply houses are recommended. Students may be given the assignment of finding their own samples of mycorrhizae by digging around the root area of pine seedlings and clipping of several root hairs. These hairs can be examined using a stereomicroscope or the low power objective lens of a microscope.

OBJECTIVE:

A. Observe mycorrhizal associations in orchids and in pine seedlings.

REQUIREMENTS FOR EACH LABORATORY SECTION

	QUANTITY	COMMENTS
Slides with mycorrhizae	1/4S	Carolina #97-7870
Root hairs with mycorrhizae	1/4S	
Stereomicroscope	1/2S	Or compound microscope
Phloxine*	1/4S	In brown bottle with dropper
Glass slides	2/S	To be provided by student
Scalpels and forceps	1/2S	

*See Appendix for formulation

EXERCISE 66
BIOINSECTICIDES

Exercise 66 introduces the student to an assay for the insecticidal properties of a biological insecticide produced by Bacillus thuringiensis var. israelensis. This organism is sold under the trade name of Dipel by Abbot Laboratories. Other companies manufacture similar preparations. As an alternative, a heavy suspension consisting of cells and endospores in TSB can be used as the source of insecticide. Students should be encouraged to examine cultures or preparations of B. thuringiensis for the presence of endospores AND parasporal bodies. Compare these with B. cereus (if such cultures are available). Examination of dead and dying larvae should also be performed. Students may work in small groups and share their results with the rest of the lab section.

OBJECTIVE:

A. Demonstrate the insecticidal activity of Bacillus thuringiensis.

REQUIREMENTS FOR EACH LABORATORY SECTION

	QUANTITY	COMMENTS
Leaves	1/S	Cabbage, broccoli, or bean.
Trichoplusia ni	10/4S	First instar larvae. May obtain from regional agricultural stations or pest control agencies. Biological supply houses (see Appendix in this manual) may also sell the larvae.
Weighing bottle	1/S	Low, flat form, 50x30mm
Bacillus thuringiensis	4/lab	Dipel, or other preparation

EXERCISE 67
EXAMINATION OF RUMEN FLUID

Exercise 67 studies some of the microorganisms and activities found in the rumen. The cellulolytic activity of these organisms will be illustrated by culturing rumen microorganisms in culture media with filter paper strips as the source of cellulose. Rumen fluid can be obtained from some agricultural stations of Academic Departments conducting investigations in the biology of ruminants or have fistulated cows (or other ruminant).

OBJECTIVE:

A.   Observe some of the rumen microorganisms;

B.   Demonstrate the cellulolytic activity of rumen microorganisms.

REQUIREMENTS FOR EACH LABORATORY SECTION

	QUANTITY	COMMENTS
Rumen fluid	10ml/lab	
Minimal-0.1% glucose broth*	1/S	10 ml in 18x150 s/c tubes with filter paper strips
Depression slides	1/S	To be provided by student
Coverslips	1/S	To be provided by student
Vaseline	1 jar	With toothpicks
GasPak jars and envelopes	1/8S	See exercise 19
Incubator	1/lab	Set at 35° to 37°C
Gram staining reagents	1/4S	

*See appendix for formulation

PART XVI
MEDICAL MICROBIOLOGY

EXERCISE 68
NORMAL FLORA OF THE BODY

Exercise 68 introduces the student to some selective procedures for the isolation of some microbial groups associated with the human body. The techniques are essentially those outlined in exercise 7E, dealing with selective and differential media. Each student should perform this important exercise and his/her results shared with others in the class at the completion of the second period.

OBJECTIVE:

A. Isolate and partially characterize resident microorganisms from various anatomical sites of the human body.

REQUIREMENTS FOR EACH LABORATORY SECTION

	QUANTITY	COMMENTS
Exercise 68A:		

Cultures
  Streptococcus
    mutans
	1/4S	48-hour in blood agar.
		ATCC 25175

Streptococcus mitis		ATCC 15909
Streptococcus salivarius		ATCC 9759
Mitis-Salivarius agar	2/S	Prepoured in plates
GasPak jars and envelopes	1/8S	See exercise 19
3% hydrogen peroxide	1/4S	In brown bottle with dropper
Gram staining reagents	1/4S	
Toothpicks, sterile	1/S	
Incubator	1/lab	Set at 35°C
Exercise 68B:		

Media
  Blood agar plates,
    5% SRBC*
	1/S	Prepoured in plates

Mannitol-salts agar*	1/S	20 ml in 18x150 tubes
Cotton (or Dacron) swabs	1/S	
Sterile saline	1/S	3ml in 13x100 s/c tubes
Incubator	1/lab	Set at 35°C

*See appendix for formulation

EXERCISE 69
ANTIBIOTIC SENSITIVITY TESTING
USING THE KIRBY-BAUER PROCEDURE

Exercise 69 introduces the student to a common procedure used in clinical laboratories to determine the susceptibility of microorganisms to antimicrobials. The results obtained using the Kirby-Bauer procedure can be compared with the MIC procedure (Exercise 22C) by using the same antibiotics, in disk form and in solution. Each student may perform the Kirby-Bauer procedure on one organism and then share the results with others in his/her group or with the rest of the class.

OBJECTIVE:

A. To determine the susceptibilities of microorganisms to antimicrobials using the Kirby-Bauer procedure.

REQUIREMENTS FOR EACH LABORATORY SECTION

	QUANTITY	COMMENTS

Cultures
  Staphylococcus
    aureus
  Escherichia coli
  Pseudomonas
    aeruginosa
	1/4S	8-hour cultures in TSB

Miller-Hinton agar*	1/S	30 ml in 100 ml bottles
15x100mm Petri plates	1/S	
Dacron-tipped swabs	1/S	
Forceps	1/2S	In alcohol jars
Antibiotic sensi- tivity disks	1 each/S	See table 69-1 for selection. May purchase from Difco or BBL (see Appendix for address).
Calipers or ruler (in mm)	1/S	
Incubator	1/lab	Set at 35°C

*See Appendix for formulation

EXERCISE 70
ABO BLOOD TYPING

Exercise 70 involves determining the ABO blood type of each student using a direct hemagglutination test. This exercise can be performed along with serological procedures outlined in exercise 71. Students may work in pairs, each lancing each other's finger. The ear lobe may be an alternate site for lancing. Since some students might "pass out at the sight of blood," they should be sitting down while they are being lanced.

OBJECTIVES:

A. To determine the ABO blood type of the student;

B. To illustrate a use for serological reactions.

REQUIREMENTS FOR EACH LABORATORY SECTION

	QUANTITY	COMMENTS
ABO blood typing kit	1/lab	Carolina* #70-4057
Bandages	1S	
or		
ABO blood typing antisera	1 each	Carolina #70-4045
70% isopropyl alcohol	100 ml	
Sterile lancets	1/S	
Cotton balls	1/S	
Bandages	1/S	
Grease marking pencil	1/S	To be provided by student
Clean glass slides	1/S	
Grease-cutting detergent	1 bar	
Toothpicks	3/6S	

*See Appendix for address

EXERCISE 71
SEROLOGICAL REACTIONS

Exercise 71 introduces the student to serological procedures commonly performed in the clinical laboratory. Each (or all) of the procedures outlined in this exercise can be performed during a 2-3 hour laboratory by small groups of students. Students should be referred to the Appendix for the proper dilution procedures and to exercise 4C for the principles and use of the fluorescence microscope. The results obtained should be interpreted and discussed briefly at the completion of each of the procedures.

OBJECTIVES:

A. To perform and interpret the following serological procedures:

  1. bacterial agglutination test;

  2. precipitin test;

  3. immunofluorescence test;

  4. rapid plasma reagin (RPR) test for syphilis

REQUIREMENTS FOR EACH LABORATORY SECTION

	QUANTITY	COMMENTS
Exercise 71A:		

Cultures
  Escherichia coli
  Salmonella
    typhimurium
	1/4S	Overnight on TSA slants

Salmonella O antiserum	1/8S	Polyvalent Difco* #2264-47
Spot plates	1/4S	
Phenol-saline solution	1/S	5 ml of aqueous 0.5% phenol and 0.85% NaCl in 13x100 s/c tubes. Suspend bacteria to a MacFarland #1 turbidity.
Pasteur pipets with bulbs	1/S	
Beaker with disinfectant	1/4S	
Exercise 71B:		
Bovine serum albumin (BSA)	1/4S	2 ml 1:500 dilution in s/c vials. May purchase BSA from Microbiological Associates*
Goat anti BSA	1/4S	May purchase from Microbiological Associates*
Normal goat serum	1/4S	May purchase from Microbiological Associates*
Normal (0.85%) saline	100 ml	
10x75mm tubes	5/S	
Pasteur pipets with bulbs	2/S	
1 ml pipets with bulbs	5/S	
Test tube rack	1/2S	
Incubator	1/lab	Set at 35°C

REQUIREMENTS FOR EACH LABORATORY SECTION

	QUANTITY	COMMENTS
Exercise 71C:		
Fluorescence microscope	1	
Cultures <u>Streptococcus faecalis</u> <u>Staphylococcus aureus</u>	1/4S	48-hour cultures in TSB
FITC-labelled anti Group D	1/8S	Difco* #2321
Gram staining reagents	1/4S	
Phosphate-buffered saline#	100 ml	pH 7.2
50% glycerol in PBS	1/8S	In bottles with dropper
Petri dishes	1/S	With a moistened filter paper
Glass slides and coverslips	1/S	To be provided by student
Pasteur pipets with bulbs	2/S	
Applicator sticks	2/S	Or toothpicks
China marker	1/S	To be provided by student
Incubator	1/lab	Set at 35°C
Exercise 71D:		
Reactive (4+) serum	1/4S	1 ml in 10x75mm tubes. May purchase from BBL* or from local health department. (Careful, such sera may also contain hepatitis or other viral pathogens.)
Non-active (-) serum	1/4S	1 ml in 10x75mm tubes

REQUIREMENTS FOR EACH LABORATORY SECTION

	QUANTITY	COMMENTS
RPR (rapid plasma reagin) test kit	1/class	Available from Hynson, Wescott & Dunning, a division of BBL*
Rotating apparatus	1/lab	Optional equipment

*See Appendix for addresses
#See Appendix in lab manual for formulation

EXERCISE 72
RAPID MULTITEST PROCEDURES FOR THE IDENTIFICATION OF ENTEROBACTERIACEAE

Exercise 72 uses the Enterotube II and the API 20E strip to illustrate to the student some of the modern multitest systems for the rapid identification of bacteria commonly encountered in the clinical laboratory. The student should be referred to exercise 28 to compare between the conventional and the rapid multitest procedures. Students may work in groups of two, using the API 20E system and the other the Enterotube II to attempt to identify the same unknown. Conventional biochemical reactions of the unknowns should be made available to the student after completion of exercise.

OBJECTIVES:

A. Identify an unknown isolate using the API 20E strip and the Enterotube;

B. To compare with rapid multitest systems with the conventional method of bacterial identification.

REQUIREMENTS FOR EACH LABORATORY SECTION

	QUANTITY	COMMENTS
Exercise 72A:		
Unknown cultures	1/S	Overnight on TSA slants.
Enterotube II	1/S	Available from Roche*
Plates of TSA	1/S	
Gram staining reagent	1/S	
Reagents Kovac's# alpha-naphthol# 20% KOH 0.3% Creatine solution Oxidase reagent#	1/4S	In bottles
Platinum inoculating loop	1/4S	
Filter paper squares	1/S	
1 ml syringes with needles	4/lab	
Incubator	1/lab	Set at 35°C
Exercise 72B:		
Unknown cultures	1/S	Overnight on TSA slants
API 20E strip	1/S	Available from Analytab*
Plates of TSA	1/S	
5ml 0.85% sterile saline	1/S	In 13x100 s/c tubes
Pasteur pipets	3/S	
Sterile mineral oil	4/lab	In bottles with dropper
Gram staining reagents	1/S	
Reagents Kovac's# 10% Ferrich chloride Sulfanilic acid# alpha-naphthylamine#	1/4S	In bottles

REQUIREMENTS FOR EACH LABORATORY SECTION

	QUANTITY	COMMENTS
alpha naphthol#		
40% KOH		
Oxidase reagent#		
Platinum inoculating loop	1/4S	
Filter paper squares	1/S	
Incubator	1/lab	Set at 35°C

*See Appendix for addresses
#See Appendix in lab manual for formulation

SOURCES OF AUDIOVISUAL MATERIALS

Abbot Laboratories
Film Service Department
North Chicago, IL 60605

American Chemical Society
1155 - 16th Street, N.W.
Washington, D. C. 20036

American Edwards Laboratories
17221 Red Hill Avenue
Irvine, Ca 92715

American Hospital Association
840 North Lake Shore Drive
Chicago, IL 60611

American Medical Association
535 N. Dearborn Street
Chicago, IL 60610

American Nurses Association
2420 Perishing Road
Kansas City, MO 64108

American Society for
   Microbiology
1913 "I" Street, N.W.
Washington, D. C. 20006

American Society of Clinical
   Pathologists
2100 W. Harrison
Chicago, IL 60612

Armed Forces Institute of
   Pathology
AudioVisual Support Center
Washington, D. C. 20305

Armstrong Industries
3660 Commercial Avenue
Northbrook, IL 60062

Association Films/Geigy
   Pharmaceutical
7838 San Fernando Road
Sun Valley, CA 91352

Audio Visual Medical
   Marketing, Inc.
850 Third Avenue
New York, NY 10022

Ayerst Laboratories, Association-
   Sterling Films
600 Garden Avenue
Ridgefield, NJ 07657

Barr Films
P. O. Box 5667
Pasadena, CA 91107

Batten, Batten, Hudson and Swab
820 Keo Way
Des Moines, IA 50309

Bausch & Lomb
635 St. Paul Street
Rochester, NY 14602

BBC TV/Open University/The Media
   Guild
11526 Sorrento Valley Road,
   Suite J
San Diego, CA 92121

BFA Educational Media
2211 Michigan Avenue
Santa Monica, CA 90406

Bio Service Corporation
500 S. Racine Avenue, Room 302
Chicago, IL 60607

Biology Media
2437 Durant Avenue, Suite 206
Berkeley, CA 94704

Bluestone Video Makers
4018 - 22nd Street
San Francisco, CA 94114

BMA Audio Cassettes
200 Park Avenue
New York, NY 10003

BNA Communications, Inc.
9401 Decoverly Hall Road
Rockville, MD 20850

Carolina Biological Supply Company
2700 York Road
Burlington, NC 27215

Carousel Films & Video, Inc.
1501 Broadway
New York, NY 10036

Center for Disease Control
U.S. Department of Health,
   Education and Welfare
Atlanta, GA 30322

Centron Films
1621 W. 9th Street
Lawrence, KS 66044

Charles Pfizer and Co.
630 Flushing Avenue
Brooklyn, NY 10006

Churchill Films
662 N. Robertson Boulevard
Los Angeles, CA 90069

CIBA Pharmaceutical Co.
Audio-Visual Department
Summit, NJ 07901

Clay Adams, Division of Becton-
   Dickinson and Co.
299 Webro Road
Parsippany, NJ 07504

Communications Skill Corporation
50 Sanford Street
Fairfield, CT 06430

Communications World
2316 - 2nd Avenue
Seattle, WA 98121

Coronet Films
65 East S. Water St.
Chicago, IL 60601

Davis and Geck Film Library,
   American Cyanamid Co.
1 Casper Street
Danbury, CT 06810

Dayton Lab Services
3235 Dayton Avenue
Lorain, OH 44055

Eastman Kodak Company
343 State Street
Rochester, NY 14650

Educational Graphic Aids, Inc.
1315 Norwood Avenue
Boulder, CO 80302

Eli Lilly and Co., Educational
   Resources Library
P. O. Box 100B
Indianapolis, IN 46206

Elliot Scientific
185 E. 85th Street
New York, NY 10028

Encyclopedia Britannica
   Educational Corporation
2494 Teagarden Street
San Leandro, CA 94577

Filmakers Library, Inc.
133 E. 58th Street, Suite 703A
New York, NY 10022

Films Incorporated
733 Green Bay Road
Wilmette, IL 60091

Fisher Scientific Company,
   Educational Materials Div.
4901 W. Le Moyne Avenue
Chicago, IL 60651

Florida State University
Instructional Media Center
Tallahassee, FL 32306

Harper & Row Audio-Visuals
2350 Virginia Ave.
Hagertown, MD 21740

Harwyn Medical Photographers
4814 Larchwood Avenue
Philadelphia, PA 19143

Human Relations Media
175 Tompkins Avenue
Pleasantville, NY 10570

IBIS Media
P. O. Box 308
Pleasantville, NY 10570

Indiana University
Audio-Visual Center
Bloomington, IN 47401

International Film Bureau, Inc.
332 South Michigan Ave.
Chicago, IL 60604

J. B. Lippincott Company
East Washington Square
Philadelphia, PA 19105

John Wiley & Sons, Inc.
605 Third Avenue
New York, NY 10016

Learning Corporation of America
1350 Avenue of the Americas
New York, NY 10019

Lederle Laboratories
Film Library
Pearl River, NY 10053

Life Sciences Associates
1 Fenimore Road
Bayport, NY 11705

McGraw-Hill Book Co.
330 W. 42nd Street
New York, NY 10036

McGraw-Hill/CRM Films
P. O. Box 641
Del Mar, CA 92014

Mecec, Inc.
12815 - 120th Avenue, N.E.
Kirkland, WA 98033

Medcom Products, Inc.
1633 Broadway
New York, NY 10019

Media Visuals
4 Midland Avenue
Hicksville, NY 11801

MEP/McGraw-Hill Publications
4530 W. 77th Street
Edina, MN 55435

Millipore Corporation
80 Asby Road
Bedford, MD 01730

Milner-Fenwick, Inc.
3800 Liberty Heights Ave.
Baltimore, MD 21215

Multi Media Publishing, Inc.
1393 S. Inca Street
Denver, CO 80223

National Audiovisual Center
General Services Administration
Washington, D. C. 20409

National Education Media, Inc.
21601 Devonshire Street, #300
Chatsworth, Ca 91311

National Health Films
P. O. Box 13973, Station K
Atlanta, GA 30324

National Medical Audiovisual
  Center
8600 Rockville Pike
Bethesda, MD 20209

Oklahoma State University
Audiovisual Center
Stillwater, OK 74078

PBS Video
475 L'Enfant Plaza, S.W.
Washington, D. C. 20024

Perennial Education, Inc.
P. O. Box 855 Rivinia
Highland Park, IL 60034

Pfizer Medical Film Library
470 Park Avenue, South
New York, NY 10016

Photography Division, Office
  of Information
U.S. Department of Agriculture
Washington, D. C. 20250

Prentice-Hall Media
150 White Plains Road
Tarrytown, NY 10591

Science Software Systems, Inc.
11899 W. Pico Blvd.
West Los Angeles, CA 90064

Scientificom
706 North Dearborn Street
Chicago, IL 60610

Shell Film Library
1433 Sadlier Circle W. Drive
Indianapolis, IN 46239

Teaching Film, Inc.
P. O. Box 66824
Houston, TX 77006

The National Foundation,
  Supply Division
P. O. Box 2000
White Plains, NY 10602

Time-Life Video
100 Eisenhower Drive
Paramus, NJ 07652

University of California,
  Extension Media Center
2223 Fulton Street
Berkeley, CA 94720

University Park Press
300 N. Charles Street
Baltimore, MD 21202

Upjohn Professional Film Library
7170 Portage Road
Kalamazoo, MI 49001

W. B. Saunders
West Washington Square
Philadelphia, PA 19105

SOURCES OF CULTURES

American Type Culture Collection
12301 Parklawn Drive
Rockville, MD   20852

Dr. Bruce Ames
Department of Biochemistry
University of California
Berkeley, CA   94720

Dr. Barbara Bachman
Curator, E. coli Genetic Stock
Department of Human Genetics
Yale University Center
School of Medicine
333 Cedar Street
P. O. Box 3333
New Haven, CT   06510

Collection Nationale d'Cultures
   de Microorganismes
Institut Pasteur
28 Rue du Docteur Roux
75724 Paris Cedex 15, France

Culture Centre of Algae and
   Protozoa
36 Storey's Way
Cambridge CB3 ODT
UK

Dico Laboratories
P. O. Box 1058A
Detroit, Mi   48201

Midwest Culture Collection
1924 North 17th Street
Terre Haute, IN   47804

National Collection of Dairy
   Organisms
National Institute for Research
   in Dairy
Shinfield, Reading RG2 9AT
UK

National Collection of Industrial
   Bacteria
Torry Research Station
135 Abbey Road
Aberdeen AB9 8DG
UK

National Collection of Type
   Cultures
Central Public Health Laboratory
Colindale Avenue
London NW9 5HT
UK

Presque Isle Cultures
P. O. Box 8191
Presque Isle, PA   16505

University Micro Reference
   Laboratory
7885 Jackson Road
Ann Arbor, MI   48103

SOURCES OF CULTURE MEDIA

Analitab Products Inc.
(source of API 20E strips)
200 Express Street
Plainview, NY 11803

BBL
Division of Becton, Dickinson
  and Co.
Cockeysville, MD 21030

Cal Labs
7332 Varna Avenue
North Hollywood, CA 91605

Carolina Biological Supply Co.
2700 York Road
Burlington, NC 27215

Carr-Scarborough Microbiologicals,
  Inc.
P. O. Box 1328
Stone Mountain, GA 30086

Difco Laboratories
Detroit, MI 48232

GIBCO
3175 Staley Road
Grand Island, NY 14072

Hardy Media
5765-B Thornwood
Goleta, CA 93117

Key Scientific Company
P. O. Box 66307
Los Angeles, CA 90066

Oxoid U.S.A. Inc.
9017 Red Branch Road
Columbia, MD 21045

Remel
Regional Media Laboratories
P. O. Box 14420
Lenexa, KS 66215

Roche Diagnostic Systems
(source of Enterotube II)
Nutley, NJ 07110

Scott Laboratories, Inc.
Fiskeville, RI

SOURCES OF SEROLOGICAL REAGENTS

BBL
Division of Becton, Dickinson
  and Co.
Cockeyesville, MD   21030

Burroughs Wellcome Co.
Wellcome Reagents Division
Research Triangle Park, NC 37709

Carolina Biological Supply Co.
2700 York Road
Burlington, NC 27215

Difco Laboratories
P. O. Box 1058A
Detroit, MI   48201

GIBCO
3175 Staley Road
Grand Island, NY   14072

Hynson, Wescott and Dunning
Division of Becton, Dickinson
  and Co.
Baltimore, MD   21201

Microbiological Associates
4733 Bethesda Avenue
Bethesda, MD   20014

Pharmacia Diagnostics
Division of Pharmacia, Inc.
800 Centennial Avenue
Piscataway, NJ   08854

SOURCES OF PREPARED SLIDES

Carolina Biological Supply Company
2700 York Road
Burlington, NC 27217

College Biological Supplies
21707 Bothell Way
Bothell, WA 98011

Fisher Scientific Company
711 Forbes Avenue
Pittsburgh, PA 15219

Macmillan Science Company, Inc.
Turtox
8200 South Hoyne Avenue
Chicago, IL 60620

Triarch, Inc.
P. O. Box 98
Ripon, WI 54971

Ward's Natural Science Establishment
P. O. Box 1712
Rochester, NY 14603

A. BACTERIOLOGY FIRST LAB EXAM

1. Explain the following terms:

    a) compound microscope
    b) simple microscope

2. What is the total magnification of a compound microscope when you use the oil immersion lens?

3. If the limit of resolution is 0.20 m, can you see the image of a bacterium that is 1.0μm by 0.10μm? Explain why.

4. What 2 characteristics of a microscope determine the resolvable distance?

    a)   b)

5. Why is oil used with the oil immersion lens (100x)?

6. What does the condenser do?

7. What is Brownian motion?

8. Name 2 types of motion found in bacteria that are due to the bacteria?

    a)   b)

9. In which kingdom (Animalia, Plantae, Fungi, Monera, Prions, Protista, Viruses) do you find the bacteria?

10. In which kingdom do you find the blue-green algae?

11. In which kingdom do you find the single-celled algae and protozoans?

12. What are 3 indicators of growth in a broth?

    a)   b)   c)

13. What type of characteristics are generally used to distinguish colonies?

    a)   b)   c)

14. List 4 genera that produce endospore.

    a)   b)
    c)   d)

15. Which 2 genera are acid-fast positive?

    a)   b)

16. What is responsible for the acid-fast characteristic?

17. What is responsible for the gram stain characteristic?

18. Draw a cross section of a gram positive and gram negative cell wall. Clearly indicate where the cytoplasmic membrane is and the cytoplasm. Label the layers of the wall clearly.

        Gram Negative   Gram Positive

19. Outline the gram stain, acid-fast stain, endospore stain, and capsule stain by listing the primary stain, mordant (if any), decolorizing agent, and counter stain in the order they are used.

    Gram Stain   Acid-Fast Stain   Endospore Stain   Capsule Stain

20. Indicate the genus and species of the organisms that cause the following diseases:

    a) tuberculosis   b) anthrax   c) botulism

21. Flagella that cover the cell are known as _____ flagella.

22. Which bacteria have axial filaments?

23. Name the bacterial genus that lacks a cell wall.

24. What are the 2 polymers found in the walls of gram positive bacteria?

    a)   b)

25. Which type of bacteria have an outer membrane in addition to their plasma membrane?

26. What color does an old gram positive organism stain? Assume its walls have broken in many places.

27. Bacterial capsules have 2 major functions. What are they?

    a)   b)

28. What does sepsis mean?

29. What are aseptic techniques?

30. What does TSA mean?

31. How do you generally sterilize:

   a) plastic petri dishes
   b) heat-sensitive liquid media

A. BACTERIOLOGY SECOND LAB EXAM

1. Match the following terms with the descriptions below:
   (catalase, oxidase, peroxidase, permease, tryptophanase, cysteine oxidase, decarboxylase).

   a) An enzyme involved in the formation of indole.
   b) An enzyme involved in the formation of H_2S.
   c) An enzyme involved in the breakdown of H_2O_2 to H_2O and O_2.
   d) An enzyme involved in the formation of an amine and CO_2 from an amino acid.
   e) An enzyme involved in the breakdown of H_2O_2 to H_2O (no oxygen is produced).
   f) An enzyme that is part of an electron transport system.
   g) An enzyme that transports citrate into a cell.

2. a) What group of organisms are responsible for fermenting cabbage to sauerkraut.
   b) Name (give the genera) 2 organisms that are responsible for the formation of sauerkraut.

3. What kind (what group) of organisms are selected for on EMB.

4. What is EMB (what do the letters stand for)?

5. Explain what a selective medium is.

6. Explain what a differential medium is.

7. Define a lactic acid bacterium.

8. Define a coliform.

9. List 5 different types of fermentation and give an example of an organism that carries out the fermentation.

Fermentation	Genus
a)	
b)	
c)	
d)	
e)	

10. List at least 4 differences between a respiration and a fermentation.

Fermentation	Respiration
a)	
b)	
c)	
d)	
e)	

11. Explain what the following enzymes do:

   a) amylase
   b) caseinase
   c) gelatinase

A.   BACTERIOLOGY THIRD LAB EXAM

1. $NH_3 \xrightarrow{\quad A \quad} NO_2^{\ominus} \xrightarrow{\quad B \quad} NO_3^{\ominus}$

   a)   What is the name of this process?
   b)   Which bacterial genus carries out A?
   c)   Which bacterial genus carries out B?
   d)   What are these bacteria called? Or what is a general name
        for these bacteria?

2. $NO_3^{\ominus} \xrightarrow{\quad C \quad} NO_2^{\ominus} \xrightarrow{\quad D \quad} N_2$

   a)   What is the name of this process?
   b)   Which bacterial genus carries out C?
   c)   Which bacterial genus carries out D?

3. $N_2 \xrightarrow{\quad E \quad} NH_3 \longrightarrow NH_2\text{-} CHR - CO_2H$

   a)   What is the name of this process?
   b)   Name 2 bacterial genera that are able to carry out E.

4. $NH_2 - CHR - CO_2H \longrightarrow NH_3$

   a)   What is the name of this process?

5.   Explain what oxidase is.

6.   What type of bacteria could have oxidase? Mark all
     possibilities.

   a)   obligate aerobes   b) obligate anaerobes   c) anaerobes
   d)   facultative anaerobes

7.   What does an oxidase positive colony look like after adding
     oxidase reagent?

8.   Explain with a chemical formula what peroxidase does.

9.   Explain with a chemical formula what catalase does.

10. If a bacterium lacks both catalase and oxidase, what kind of an
     organism is it?

11. List $\underline{3}$ important characteristics of lactic acid bacteria.

   a)
   b)
   c)

12. Name $\underline{4}$ genera of lactic acid bacteria.

13. Name 2 tests you can easily carry out on a bacterium to determine whether or not it is a lactic acid bacterium.

    a)   b)

14. List 3 important characteristics of coliforms.

    a)
    b)
    c)

15. Name 4 genera of coliforms.

16. Which genus is generally used as an indicator of fecal contamination of water and foods?

17. List, in order, the 3 parts of the standard analysis of H O and indicate the name of the medium and the positive results for E. coli.

Test	Medium	Positive Results
a)		
b)		
c)		

18. Explain the difference between pasteurization and sterilization.

19. Indicate the genus and species of the organisms that cause the following diseases:

    a) Brucellosis   b) Q-fever
    c) Mastitis   d) Tuberculosis in cattle

20. Indicate the genus and species of the organisms that cause the following diseases (consider only bacteria):

    a) Typhoid fever   b) Cholera
    c) Gastroenteritis   d) Dysentery

21. Diseased cattle can give rise to milk contaminated with which microorganisms (genus is sufficient):

    a)   b)
    c)   d)

22. Fecal contamination of milk often results in the presence of which microorganisms (genus is sufficient):

    a)   b)   c)

A. BACTERIOLOGY LAB FINAL

1-5. Using a five kingdom system of classification (a = protista, b = plantae, c = animalia, d = prokaryotae, e = fungi), in which kingdom would you place each of the following organisms:

    1. bacteria
    2. yeast
    3. protozoans
    4. blue-green algae
    5. single-celled eukaryotic algae

6. What is the diameter of the average bacterium?

    a) 1 mm, b) 100 μm, c) 1 μm, d) 10 nm, e) 1 nm

7. What is the approximate resolution of the best light microscopes?

    a) 0.2 mm, b) 0.2 μm, c) 0.2 nm, d) 0.2 angstrom, e) none of the above.

8. The reason you cannot see any of the viruses with the light microscope is because one or more of their dimensions is smaller than 1000 angstroms.

    a) true, b) false

9. A basic dye is one where the colored portion is negatively charged.

    a) true, b) false

10. Bacteria generally have a great number of negatively charged groups associated with their cell-walls, cell-membranes, and in their cytoplasm and so are readily stained with basic dyes.

    a) true, b) false

11. Which bacteria are catalase negative?

    a) <u>Escherichia</u> and <u>Klebsiella</u>, b) <u>Streptococcus</u> and <u>Lactobacillus</u>, c) <u>Mycobacterium</u> and <u>Nocardia</u>, d) <u>Bacillus</u> and <u>Thermoactinomyces</u>, e) all of the above.

12. The gram stain tells you the approximate structure and chemical composition of the cytoplasmic membrane.

    a) true, b) false

13. The gram stain on an old gram positive bacterium may color the
bacterium pink (red).

    a) true, b) false

14. An acid-fast positive bacterium stains blue after the acid-fast
stain.

    a) true, b) false

15. After an endospore stain on Mycobacterium, the vegetative cells
stain green.

    a) true, b) false

16. After an endospore stain on Bacillus and Clostridium, the
vegetative cells stain green.

    a) true, b) false

17. Which bacterium does not form endospores?

    a) Sporolactobacillus, b) Desulfotomaculum, c) Sporosarcina,
    d) Lactobacillus, e) Bacillus.

18. Which bacteria are acid-fast positive?

    a) Nocardia and Mycobacterium, b) Bacillus and Clostridium,
    c) Escherichia and Salmonella, d) Microbacterium and
    Mycobacterium, e) none of the above.

19. Gram positive bacteria generally have a much thicker
peptidoglycan layer in their cell wall than gram negative
bacteria.

    a) true, b) false

20. Gram negative bacteria have an outer membrane which is
considered to be part of the cell-wall.

    a) true, b) false

21. Gram positive bacteria have a large amount of teichoic acid
associated with their cell-walls while gram negative bacteria
have none.

    a) true, b) false

22. Acid-fast positive bacteria are acid-fast positive because
a) they have many negatively charged groups associated with
their wall, b) they have a very thick cell-wall, c) they have
an outer membrane associated with their cell-wall, d) they
have a thick layer of waxes and lipids associated with their
cell-wall, e) none of the above.

23. Vegetative cells are generally killed within seconds when exposed to boiling water.

    a) true, b) false

24. What kind of organisms can you enrich (select) for by heating a soil at 80° C for 15 to 20 minutes?

    a) acid-fast positive bacteria, b) lactic acid bacteria,
    c) nitrifying bacteria, d) endospore-forming bacteria,
    e) none of the above.

25. Soil can easily be sterilized by boiling it in water for 2 or 3 minutes.

    a) true, b) false

26. Freezing food or water is a very effective way of sterilizing it.

    a) true, b) false

27. Vegetative growth is said to occur when bacteria grow and divide.

    a) true, b) false

28. Sporulation is said to occur when endospores develop back into bacteria.

    a) true, b) false

29. Germination is said to occur when a vegetative cell develops into an endospore.

    a) true, b) false

30. Bacteria in the genus Streptomyces are known to produce spores.

    a) true, b) false

31. Bacteria in the genus Streptomyces are known to produce many different antibiotics.

    a) true, b) false

32. Ultraviolet light is used for sterilizing the air and the surfaces of working areas.

    a) true, b) false

33. Most media and equipment such as glass pipets and glass petri
    dishes are sterilized by steam under pressure.

    a) true, b) false

34. Heat labile (sensitive) media are often sterilized by
    filtration while heat labile equipment is sterilized by ultra-
    violet light or with ethylene oxide gas.

    a) true, b) false

35. If a broth tube shows no turbidity, we can conclude that no
    growth has occurred.

    a) true, b) false

36. Which are indicators of growth in a broth tube?

    a) pellicle formation, b) turbidity, c) sediment, d) all of
    the above, e) b and c only.

37. When you make sauerkraut, you generally have to add bacteria to
    get the correct fermentation started.

    a) true, b) false

38. When you make sauerkraut, you add enough NaCl to the shredded
    cabbage so that you have a concentration of salt close to 2.5%
    (w/w).

    a) true, b) false

39. Which organisms proliferate in fermenting cabbage as it
    develops into sauerkraut?

    a) acid-fast bacteria, b) coliforms, c) oxidase negative
    bacteria, d) yeast, e) lactic acid bacteria.

40. Which organisms are most prevalent on cabbage before
    fermentation begins?

    a) denitrifying bacteria, b) lactic acid bacteria,
    c) nitrifying bacteria, d) coliforms and other gram negative
    bacteria, e) none of the above.

41. When you make yogurt, you generally have to add bacteria to get
    the correct fermentation started.

    a) true, b) false

42. The bacteria you generally find in yogurt are:

a) Micrococcus and Staphylococcus, b) Neisseria and Chlamydia, c) Corynebacterium and Nocardia, d) Lactobacillus and Streptococcus, e) none of the above.

43. In order to get the yogurt bacteria to grow well, you must incubate them at around 45° C.

a) true, b) false

44. Different strains of Saccharomyces cerevisiae are used to make wines and beers.

a) true, b) false

45. Saccharomyces cerevisiae is used by bakers to make bread rise.

a) true, b) false

46. What kind of an organism is Saccharomyces?

a) bacterium, b) virus, c) protozoan, d) fungus, e) none of the above

47. What kind of fermentation is occurring when you make good wine or beer?

a) lactic acid ferment, b) mixed acid ferment, c) 2,3-butanediol ferment, d) alcoholic ferment, e) none of the above

48-52. Fermentation tubes.

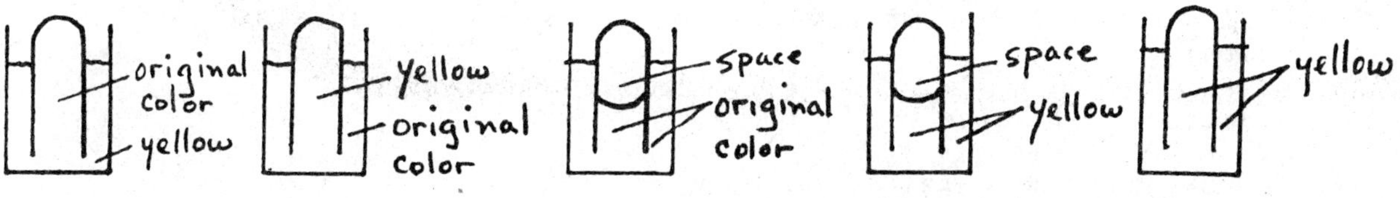

48. In a pure alcoholic fermentation, how would the fermentation tube appear?

49. In a pure lactic acid fermentation, how would the fermentation tube appear?

50. In a mixed acid fermentation, how would the fermentation tube appear?

51. An organism both ferments and respires in a fermentation tube. When it respires, NH_3 is released. How would the fermentation tube appear?

52. An organism wich has its maximum growth rate at a temperature between $0°C$ and $15°C$ is called a:

    a) thermophile, b) mesophile, c) basophil, d) psychrophile,
    e) none of the above

53. The greening and slight clearing around colonies on a blood agar plate is referred to as a:

    a) gamma reaction, b) beta hemolysis, c) alpha hemolysis,
    d) all of the above, e) none of the above

54. Which of the following organisms are sometimes found in unpasteurized milk from diseased animals?

    a) <u>Mycobacterium</u>, b) <u>Coxiella</u>, c) <u>Brucella</u>,
    d) <u>Staphylococcus</u>, e) all of the above

55. The methyl red test (if positive) tells you that your bacterium produces a lot of acid.

    a) true, b) false

56. The Voges-Proskauer test (if positive) tells you that your bacterium produces acetylmethylcarbinol (acetoin) and therefore carries out the 2,3-butanediol fermentation.

    a) true, b) false

57. There are bacteria that are negative for both the Voges-Proskauer test and the methyl red test.

    a) true, b) false

58. All aerobic organisms have either peroxidase or catalase (or both).

    a) true, b) false

59. Peroxidase and catalase destroy

    a) antibiotics, b) superoxide (O_2^-), c) toxic alcohols,
    d) peroxide (H_2O_2), e) none of the above.

60. Peroxidase is a cytochrome found in some electron transport systems.

    a) true, b) false

61. Oxidase is a cytochrome found in some electron transport systems.

    a) true, b) false

62. A bacterium which lacks both peroxidase and catalase would be

a) an aerobe like Micrococcus, b) a facultative anaerobe like Escherichia, c) an obligate anaerobe like Clostridium, d) an anaerobe which is tolerant to oxygen like Streptococcus, e) none of the above

63. A bacterium which lacks oxidase could be

a) an aerobe, b) a facultative anaerobe, c) an obligate anaerobe, d) all of the above, e) none of the above

64-67. Indicate a genus of bacterium

a) Escherichia, b) Salmonella, c) Shigella, d) Vibrio, e) Enterobacter

that contains a species which is responsible for the diseases below.

64. cholera

65. typhoid fever

66. dysentery

67. Which organism causes root nodules (where nitrogen fixation occurs) in legumes?

a) Agrobacterium, b) Clostridium, c) Rhizobium, d) Azotobacter, e) Nitrosomonas

68. Which bacterium is usually considered an indicator of fecal contamination of water or food?

a) Escherichia, Salmonella, c) Shigella, d) Vibrio, e) Enterobacter

69. Water is put into 2x lactose fermentation tubes and then the tubes are incubated for 24 hours. If the tubes turn yellow and there is a lot of gas in the Durham tubes, this is presumptive evidence for which bacterium?

a) Salmonella, b) Shigella, c) Escherichia, d) Vibrio, e) all of the above

70. If the presumptive test (previous question) is positive for acid and gas, what organisms could be present?

a) Klebsiella, b) Enterobacter, c) Citrobacter, d) Escherichia, e) all of the above

71. To test for the presence of Escherichia in 2x lactose fermentation tubes which indicate acid and gas production, you would inoculate brilliant green bile lactose (BGBL) broth with material from the 2x lactose fermentation tubes. What is a positive test for E. coli in BGBL broth tubes?

    a) green color but not gas, b) yellow color and gas,
    c) yellow color but no gas, d) green color and gas, e) none of the above

72. The test described in the previous question (involving BGBL broth) is known as the completed test.

    a) true, b) false

73. Which bacteria are involved in nitrogen fixation?

    a) Azotobacter, b) Rhizobium, c) Clostridium,
    d) Anabaena, e) all of the above

74. Which bacteria are involved in denitrification?

    a) Pseudomonas, b) Nitrosomonas, c) Nitrobacter,
    d) all of the above, e) none of the above

75. Which bacteria are involved in nitrification?

    a) Azotobacter, b) Rhizobium, c) Clostridium,
    d) all of the above, e) none of the above

76. Name the process: $NO_3^- \text{ -------} > N_2$.

    a) nitrogen fixation, b) denitrification, c) nitrification,
    d) ammonification, e) none of the above

77. Name the process: $N_2 \text{ ------} > NH_2CHOOH$.

    a) nitrogen fixation, b) denitrification, c) nitrification,
    d) ammonification, e) none of the above

78. Name the process: $NH_3 \text{ -------} > NO_3^-$.

    a) nitrogen fixation, b) denitrification, c) nitrification,
    d) ammonification, e) none of the above

79-82. Mark a) if the answer given next to the number is correct, but mark b) if the answer is incorrect.

Number per ml in the tubes.
79. 1,000,000
80. 10,000
81. 100
82. 10

Number on the plate.
83. 1

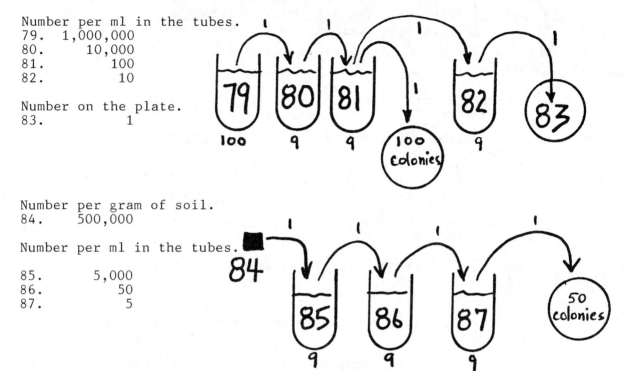

Number per gram of soil.
84. 500,000

Number per ml in the tubes.

85. 5,000
86. 50
87. 5

88. Which medium would be best to select for and and detect coliforms?

a) plate count agar (PCA), b) sabouraud (SAB), c) milk agar, d) trypticase soy agar (TSA), e) eosin methylene blue (EMB)

89. Which medium is used to select for yeast and filamentous fungi?

a) PCA, b) SAB, c) milk agar, d) TSA, e) EMB

90. What is responsible for the destruction of tissue by beta-hemolytic Streptococcus?

a) complement, b) antibodies, c) antigens, d) antibiotics, e) exoenzymes

91. Which disease is due to cross-reacting antibodies made against a Streptococcus?

a) mastitis, b) strep throat, c) scarlet fever, d) rheumatic fever, e) tooth decay

92. In humans, which organisms are generally found in the throat?

    a) Saccharomyces, b) Penicillium, c) Entamoeba,
    d) Streptococcus, e) Giardia

93. Which organisms are generally found on the skin of humans and animals?

    a) Lactobacillus, b) Pediococcus, c) Escherichia,
    d) Staphylococcus, e) Streptococcus

94. Which organisms are responsible for crown gall?

    a) Rhizobium, b) tobacco mosaic virus, c) Erwinia,
    d) Agrobacterium, e) Streptococcus

95. Which ingredient is not found in a defined medium?

    a) KH_2PO_4, b) $(NH_4)2SO_4$, c) yeast extract, d) $MgSO_4$,
    e) trace elements

96. A material that is pasteurized is sterile?

    a) true, b) false

97. Milk is pasteurized in order to protect us from diseases such as

    a) dysentery, b) tuberculosis, c) Q-fever, d) undulant fever, e) all of the above

98. The bacteria which generally spoil hamburger are the

    a) lactic acid bacteria, b) proteolytic and lipolytic bacteria, c) coliforms, d) the nitrogen fixing bacteria,
    e) nitrifying bacteria

99. All bacteria grow best at a temperature around 35°C and this is why all bacteriology labs have incubators set at 35°C.

    a) true, b) false

100. Agar can be used by most bacteria as a carbon and energy source.

    a) true, b) false.

B.   BACTERIOLOGY FIRST LAB EXAM

FILL-INS: Correctly complete the following statements.

1. Three broad groups of microoranisms are eucaryotic - namely, protozoa, _____, and algae.

2. Of the five broad groups of microorganisms, two consist entirely of photosynthetic species: _____ and algae.

3. The function of the iris diaphragm of a compound light microscope is to _____.

4. Resolution of a light microscope is definable as _____ _____.

5. The total magnification of a compound light microscope when an eyepiece of 10X and an oil immersion objective of 90X power are in use would be _____.

6. When one detects the presence of a minute object such as a bacterial cell on a slide by means of darkfield microscopy, what will be the appearance of the image? _____ _____.

7. The limit of resolution of the compound light microscope under the best possible conditions of use is approximately _____ _____.

8. Distinguish between true motility and Brownian movement as to their mechanism (cause). _____ _____.

9. Even if "perfect" staining technique is practiced in the gram stain a gram-positive organism may stain as if it is gram-negative if the culture under study is too _____.

10. The names of the gram stain reagents (other than water) in their proper sequence of use is: _____, _____, _____, _____.

11. What is a definition (or reasonable description of a mordant? _____.

12. At neutral pH, most useful bacteriological dyes are _____ charged in aqueous solution.

13. Acid-fast bacteria belong to the genus? and would stain? in the hot acid-fast staining technique: genus _____, color _____.

14. An example of a negative staining technique would be the
_____ that you performed in class, because _____
_____ .

15. The limit of resolution of a typical light microscope is:
_____

16. What is the total magnification obtainable with a binocular
light microscope if the eyepiece lenses are 15X and the
objective is 30X?

17. _____ is the name of the mordant used in the
gram stain.

18. Gram-positive cells should stain what color?

19. Give an example (genus and species) of a medically important
acid-fast bacterium.

20. Name one advantage of the hanging drop technique over an
ordinary wet mount when one is trying to detect motility.

21. Name one difference between endospores and vegetative cells of
the same species, other than shape or staining reactions.

22. What is the usual chemical composition of bacterial capsules?

23. Why is Maneval's stain technique for detection of capsules
considered an example of a "negative" staining technique (i.e.,
distinguish between "positive" and "negative" staining
techniques)?

24. What is the purpose of using aseptic technique in the
manipulaton of pure cultures of bacteria?

25. Give a characteristic of agar which is obviously advantageous
for a solidification agent of culture media.

26. Where would you expect to find a pellicle?

27. What is the basic purpose for using a special technique to
streak a petri plate medium?

28. Describe how you would distinguish between a water-soluble and
a water-insoluble bacterial pigment.

29. Name two distinct methods of sterilization and give an example
of how each method is used in bacteriology/microbiology.

   a.
   b.

30. What would be the probable result if you performed a gram stain on a mixed smear of <u>Staphylococcus</u> <u>epidermidis</u> and <u>Escherichia</u> <u>coli</u>, but:

   a. decolorized the smear with acid-alcohol rather than alcohol?

   b. decolorized with alcohol, but for too brief a time?

MULTIPLE CHOICE: Circle the letter in front of each entry which accurately answers/completes/describes the preceding numbered statement. More than one entry may be correct for a given statement.

31. Procaryotic groups of microorganisms include:

   a. bacteria
   b. fungi
   c. blue-greens

32. It is possible to detect:

   a. bacterial flagella in hanging drop preparations
   b. bacterial endospores in stained, fixed preparations
   c. viruses with the light microscope

33. On the light microscopes, such as are in use in this laboratory:

   a. the approximate limit of resolution, with the 100X objective in use, is 0.4 m
   b. if the iris diaphragm is closed too far, the limit of resolution will be larger (greater than theoretically achievable)
   c. total magnification, with the high-dry lens in use, is 1000X

34. In the gram stain technique:

   a. the proper sequence of reagents, other than water rinses, is crystal violet, ethanol, Gram's iodine, safranin
   b. omission of Gram's iodine will tend to cause all bacteria to stain as if they were gram negative
   c. excessive use of ethanol will tend to cause all bacteria to stain as if they were gram-negative
   d. properly stained, gram-positives appear purple and gram-negatives appear pink/red

35. In the capsule stain technique:

   a. Congo red colors the capsules red
   b. this technique is an example of a negative stain
   c. the older the culture, the more likely it is that capsule production will be detected

36. Acid-fast bacteria:

    a. include the agents of anthrax and leprosy
    b. owe their resistance to decolorization in acid-fast stain
       techniques to the presence of a large amount of protein in
       their cell walls
    c. stain deep red/crimson by the particular technique use in
       class

37. Endospores:

    a. resist the lethal effects of heat better by far than any
       other life-form we know on this planet
    b. resist the penetration of the usual basic dyes applied at
       room temperature
    c. are produced by the bacterial agents of tetanus and botulism

38. In relation to the transfer of microbial cultures and the
    control of microbial contamination in the laboratory:

    a. swabbing your lab benches before and after work with
       cultures is an example of sterilization
    b. "aseptic" means <u>without</u> <u>contamination</u> rather than
       <u>against</u> <u>contamination</u>
    c. the autoclave is commonly used to sterilize culture media
       and sterile reagents before use and to sterilize
       contaminated materials before disposal

39. What name is given to the methods of manipulation of bacterial
    cultures which are designed to avoid contamination?

40. Briefly describe the chemical properties and source of agar.

41. An isolated population of bacterial cells (somewhat) resembling
    a bit of colored paste on solid culture media is known as a

    _____.

42. Describe the typical conditions employed when using an
    autoclave to sterilize tubed culture media.

43. What are 2 of the notable differences between the vegetative
    cells and the endospores of a typical <u>Bacillus</u> species?

44. Give an example of a chemically undefined medium and explain
    why you feel that it is an appropriate example.

45. Give an example of a gaseous sterilant and an example of a
    situation or item which would <u>best</u> be sterilized by a gaseous
    method.

46. Why do you cool the melted agar pours to 45-47°C (approx. 50°C)
    <u>before</u> pouring the agar plates?

47. An example (genus and species) of an acid-fast bacterium of medical importance is _____

48. What is a peptone and what is it used for in bacteriology?

B. BACTERIOLOGY SECOND LAB EXAM

TRUE-FALSE: Circle the T or F as is most appropriate.

1. T F Microorganisms growing at 35° C are classified as thermophiles.

2. T F Bacteria are unable to grow at or below 0° C.

3. T F Fungi tend to tolerate acidic environments less well than bacteria.

4. T F Strictly aerobic bacteria will not grow in the absence of oxygen.

5. T F Organisms which grow well when the oxygen in their environment is kept at a very low level are described as microaerophilic.

6. T F Escherichia coli is an example of a strict anaerobe.

7. T F Starch hydrolysis is produced by extracellular enzymes known as peptidases.

8. T F If a nutrient gelatin culture grown at 37° C solidifies only after it is refrigerated for 30 minutes, this is evidence for the organism being ale to digest gelatin.

9. T F When a clear area develops in a skim milk agar plate culture immediately around the growth, this is evidence that the organism hydrolyzes casein.

10. T F Bacteria are able to phagocytose small food particles.

11. T F Too heavy an inoculum can cause a false positive reading of the citrate test.

12. T F Indole is a waste product from the breakdown of the amino acid methionine.

13. T F The products of lysine decarboxylation include CO_2 and the organic amine called cadaverine.

14. T F Hydrogen sulfide production from the breakdown of S-containing amino acids is detected by the addition of Kovac's reagent.

15. T F Amino acid decarboxylation reactions can result in the formation of ATP.

16. T F Growth in Kosers citrate broth indicates the organism has the ability to use citrate as its sole organic carbon source.

17. T F Failure to grow on starch agar indicates an organism is unable to digest starch.

18. T F If both tubes in the lysine decarboxylase test turn from yellow to purple after incubation, this is evidence that lysine has been decarboxylated.

19. T F Preservation of salt-cured meats against bacterial spoilage involves inhibition of microbial growth by a hypertonic environment.

20. T F Psychrophiles are microbes which grow only if the environment contains a high amount of dissolved fats.

B. BACTERIOLOGY THIRD LAB EXAM

TRUE-FALSE: Circle the T or F in front of each statement, as appropriate.

1. T F Failure of the bromcresol purple indicator in a glucose broth culture to change color following incubation may not be due to an inability of the organism to ferment glucose (in some instances).

2. T F Any sign of a yellow color in a BCP fermentation broth culture, no matter how slight, is interpreted as positive evidence of fermentation.

3. T F If an organism ferments lactose, it can be predicted that it should also ferment glucose.

4. T F The Voges-Proskauer test must be performed before the methyl red test is done on the same MRVP broth culture because the VP test requires analkaline pH to work properly.

5. T F MRVP broth is essentially a glucose peptone broth with no pH indicator added initially.

6. T F If an organism grows in MRVP broth, you should expect either or both tests (MR, VP) to be positive.

7. T F Catalase activity is present in most bacteria which grow aerobically but absent in most bacteria which grow anaerobicallly.

8. T F Evolution of gas by a bacterial culture to which hydrogen peroxide has been added is positive evidence that the culture possesses either catalase or peroxidase activity.

9. T F Hektoen enteric agar is differential for lactose-fermenting and for H_2S-producing bacteria.

10. T F All selective culture media contain specific inhibitory compounds (high salt, sugar, dyes, antibiotics, etc.) as the basis for their selectivity in allowing only certain bacteria to grow.

11. T F Starch agar was an example of a selective plating medium.

12. T F Casein (skim milk) agar was an example of a differential plating medium.

13. T F Oxidase activity is another method possessed by some bacteria for the detoxification of hydrogen peroxide.

231

14. T F Bacteria which show tryptophane deaminase activity would
       be likely to give a positive test result for
       ammonification in peptone broth culture.

15. T F The development of a distinct yellow color when peptone
       broth (or nitrate broth) cultures are mixed with
       Nessler's reagent in spot plates (yellow relative to test
       with sterile negative broth control) is evidence for the
       presence of free ammonia (as ammonium ions) in the
       culture.

16. T F Denitrification is an anerobic process performed by
       microbes which can carry out anaerobic respiration and
       use nitrate or nitrite ions as terminal electron
       acceptors in place of molecular oxygen.

17. T F The evolution of gas from a nitrate broth culture only
       after addition of sulfamic acid and zinc is positive
       evidence that the culture has reduced nitrate to nitrite.

18. T F Nitrogen fixation is performed by many blue-green algae,
       Azotobacter species, and Clostridium pasteurianum as a
       nonsymbiotic process.

19. T F Nitrogen fixation is an anerobic process, since it
       involves the reduction of molecular nitrogen to the state
       of amino groups and incorporation of such reduced
       nitrogen forms into organic matter (e.g., to form amino
       acids).

20. T F Urease activity will result in the production of one
       molecule of ammonia from each molecule of urea, plus one
       molecule of carbon dioxide.

B. BACTERIOLOGY FOURTH LAB EXAM

1. You performed a standard plate count experiment on a soil specimen and made the following observations. Ten serial 10-fold dilutions of the soil in sterile saline diluent were and 0.1 ml samples were plated, in triplicate, from the three highest dilutions on NA (nutrient agar). Plates were incubated at ambient (room) temperature for one week and the counts listed below obtained:

Dilution tube	Colony counts (3 plates)
#8	650, 490, 540
#9	43, 52, 58
#10	6, 2, 11

   a. What is the most statistically reliable set of plate count data?

   b. What is the average viable cell count/ml of diluted soil in the dilution used to make the plates for the answer to part a?

   c. What is the most reliable average viable cell cont/gm of the original soil specimen? (assume 1 gn, like 1 ml, occupies 1 cc of volume; show calculations for complete credit)

   d. Name one physiological/nutritional/environmental type of bacterium (not genus and species), which would not be detected in the soil specimen above, although that type is quite probably present and alive.

2. Why does the public health department not routinely examine samples from drinking water sources for the various pathogenic bacteria, amoebae, and viruses that might be present in polluted supplies?

3. In relation to water analysis for signs of pollution, what are coliforms (define them) and why are they sought routinely as an indicator of water pollution?

4. Name one advantage of the membrane filter technique over the conventional multiple tube method of water analysis for the detection of coliforms.

5. What would you expect to find in a "plaque," on a petri dish cultue of E. coli?

6. What is the sequence of major events that occurs when a bacterial virus infects and multiplies inside of a suitable host bacterial cell (seven major events)?

7. Name one major way in which the bacteria differ structurally from true fungi?

8. One genus of fermentative bacteria commonly found in fermented foods is: _____.

9. Name a chemotherapeutic drug that is <u>not</u> an antibiotic.

B. BACTERIOLOGY LAB FINAL

TRUE AND FALSE QUESTIONS

1. Immersion oil is used with the oil-immersion objective to help prevent loss of light due to bending of light rays.

2. When observing a hanging drop preparation, one should employ light of greater intensity than when examining stained preparations.

3. The diameter of an average bacterium is about 1 mm.

4. A basic dye is one where the colored portion of the molecule is negatively-charged.

5. The gram stain tells you the approximate structure and chemical composition of the plasma membrane.

6. After an endospore stain on <u>Mycobacterium</u> (an acid-fast bacterium), the vegetative cells stain green.

7. The resolving power of a lens system increases as the diameter of the smallest visible structure decreases.

8. With a parfocal microscope, once an object has been focused under a given objective lens, it need not be refocused with another lens.

9. Gram positive bacteria tend to become gram negative in older cultures.

10. Acid-fast stains are employed in the examination of sputa from patients suspected of having tuberculosis.

11. At temperatures of 100°C, the pressure used in the autoclave is an effective killing agent.

12. The spores formed by <u>Bacillus</u> are called zygospores.

13. Acid-fast organisms contain a high concentration of ribonucleic acids near the cell surface.

14. If the gram stain were stopped and the slide examined just after the application of the iodine, gram negative bacteria would stain pink.

15. Capsules are especially hard structures to stain and heat is generally employed.

16.   Pipettes and glass tubes are usually sterilized using filters.

17.   Endospores are generally killed at the same rate that vegetative cells are killed when exposed to boiling water.

18.   Readings of 15 psi in an autoclave, alone, are sufficient indication that the sterilization procedure is going to be successful.

19.   Most culture media are sterilized by autoclaving.

20.   If a tube of broth that thas been inoculated with a bacterium shows no turbidity, we can conclude that no growth has occurred.

21.   When you make sauerkraut, you generally have to add bacteria to get the fermentation of cabbage started.

22.   Bacteria resistant to heat by virtue of endospores are rarely isolated from garden soils.

23.   Cotton plugs are used in test tubes because they permit air and bacteria to enter the capped tubes.

24.   At concentrations of about 1.5%, agar melts and solidifies at 55°C.

25.   Nutrient gelatin contains approximately 1.5% agar.

26.   Petri plates are incubated in an inverted position to prevent the disruption of surface colonies by moisture droplets.

27.   Metchnikov's reagent is used to test for indole production.

28.   Casein hydrolysis is tested by adding gram's iodine to the plate.

29.   In order to get yogurt bacteria to grow well, you must incubate the milk at temperatures above 40°C.

30.   The anaerobes, almost without exception, produce catalase.

31.   Bacteria are eukaryotic organisms.

32.   Lactic acid fermentation is the predominant type of fermentation carried out by wine yeasts.

33.   Lactic acid bacteria may produce alcohol as well as lactic acid during their fermentation of carbohydrates.

34.   Organisms that do not ferment never turn the tubes of phenol red-glucose yellow due to acid production.

35. The methyl red test tells you that the bacterium produces a lot of acid (mixed acid fermentation).

36. The Voges-Proskauer test (if positive) tells you that your unknown carries out a 2,3-butanediol fermentation.

37. All aerobic organisms have either peroxidase or catalase.

38. Oxidase is a cytochrome found in some electron transport systems.

39. Water is put into 2x lactose broth and then the tubes (all 5 of them) are incubated at 35°C for 48 hours. If the tubes turn yellow only, that is ample indication that coliforms are present.

40. The presumptive test for coliforms involves the inoculation of BGBL broth tubes with the unknown sample of water.

41. The conversion of nitrate to molecular nitrogen is known as nitrification.

42. The conversion of molecular nitrogen to organic nitrogen is known as nitrification.

43. The conversion of organic nitrogen into ammonia is known as ammonification.

44. Nitrate reduction involves not only the conversion of nitrate to nitrite and nitrogen, but also to ammonia.

45. Agar slopes are usually employed to isolate pure cultures of microorganisms.

46. If a sample of milk is diluted 1000 times, and then 1 ml of it plated, the final dilution value is 1/10,000.

47. If the above dilution were plated and 45 colonies appear on the plate, the total count would be 450,000 cfu/ml.

48. If a plate receiving 0.2 ml of a 1/10,000 dilution yields a total of 90 colonies, the number of colonies/ml of sample (undiluted) is 1,800,000/ml.

49. If 0.5 ml of milk is mixed with 9.5 ml of water, the resulting dilution is 1/20.

50. If 0.3 ml of the above dilution were plated and 50 colonies develop on the plate, the number of microorganisms/ml of sample would be 3300.

MULTIPLE CHOICE QUESTIONS

Fill in the table below:

NAME OF TEST	MEDIUM USED	PRINCIPAL INGREDIENT	REAGENT USED	APPEARANCE OF POSITIVE RESULTS
Starch	51	52	53	54
55	56	tryptophan	57	58
59	60	61	62	medium turns black
63	64	65	methyl red	66

51. a) SIM, b) starch agar, c) glucose agar, d) casein agar

52. a) SIM, b) Simmons citrat, c) PCA, d) SAB

53. a) iodine, b) Metchnikov's, c) sulfamic acid, d) Kovak's

54. a) red, b) black, c) clear area around colony, d) bubbles

55. a) citrate, b) starch, c) hydrogen sulfide production,
     d) indole

56. a) SIM, b) Simmons citrate, c) TSA, d) PCA

57. a) Kovac's, b) Metchnikov's, c) sulfamic acid, d) alpha
     naphthol

58. a) black rink, b) yellow ring, c) red ring, d) black
     precipitate

59. a) hydrogen sulfide, b) indole, c) nitrification,
     d) casein

60. a) SIM, b) PCA, c) TSA, d) EMB

61. a) tryptophan, b) cysteine; c) lysine, d) glucose

62. a) black precipitate, b) red precipitate, c) neither one

63. a) methyl red, b) Voges-Proskauer, c) indole,
     d) decarboxylase

64. a) MRVP, b) indole, c) SIM, d) TSA

65. a) glucose, b) amino acids, c) starch, d) lysine

66. a) turns red, b) turns yellow, c) turns green, d) turns
     clear

67. The word sterile means:

    a) free of microorganisms, b) free of living microorganisms,
    c) clean, d) free of viruses, e) none of the above

68. The most common way of assessing the quality of foods like
    hamburger is by:

    a) the MPN test, b) the standard plate count, c) antibiotic
    sensitivity tests, d) the gram stain

69. Antibiotic sensitivity tests involve the use of:

    a) disinfectants and antiseptics as control agents,
    b) pennies, c) antibiotic-impregnated filter paper disks,
    d) none of the above

70. A common technique to measure antibiotic sensitivity is the:

    a) maximum inhibitory concentration procedure; b) the TSA
    procedure, c) the Muller-Hinton test, d) the Kirby-Bauer
    test

71. Hemolysis is:

    a) the breakdown of blood by hypotonic solutions, b) the
    breakdown of casein by bacteria, c) the breakdown of blood
    by microorganisms, d) the autolysis of bacterial cells

72. Fermented milk beverages are made by the action of _____
    on the milk.

    a) yeasts, b) lactic acid bacteria, c) spore-formers,
    d) molds

73. When bacteria convert ammonium in soils into nitrate, the
    process is called:

    a) nitrofication, b) denitrification, c) nitrogen fixation,
    d) putrefaction

74. The above bacteria are:

    a) chemoorganotrophs, b) chemoheterotrophs, c) chemolitho-
    trophs, d) none of these

75. The spread plate is a common technique to:

    a) isolate pure cultures of microorganisms, b) quantitate the
    number of aerobic organisms on samples, c) obtain large
    quantities of bacteria, d) none of the above.

76-100. Outline the procedure you would follow in order to isolate
    a chemoautotrophic, mesophilic bacterium from a soil sample.

C. BACTERIOLOGY FIRST LAB EXAM

1. Bacteria have a number of shapes. Give the technical term that corresponds to the following shapes:

   a) rod   b) sphere
   c) short spiral   d) long helix

2. Bacterial cells are often arranged in characteristic groupings. Indicate with a drawing how the following bacteria are arranged.

   a) sarcina   b) streptococcus
   c) diplococcus   d) staphylococcus

3. Bacteria have 2 types of appendages they use to propel themselves. What are they called?

   a)   b)

4. Explain the difference between vital motion and Brownian motion.

5. What does a gram stain tell you about a bacterium besides its gram reaction?

6. Which 2 bacterial genera are acid-fast positive?

   a)   b)

7. Name 6 bacterial genera that are endospore formers?

   d)   e)   f)

8. Which group of bacteria lack a cell-wall?

9. Capsules may be constructed from 2 types of material. What are they.

   a)   b)

10. Write the genus and species of the bacteria responsible for the following diseases.

    a) tuberculosis   b) tetanus   c) gas gangrene

    d) anthrax   e) leprosy   f) botulism

11. Name the 3 indicators of growth.

    a)   b)   241   c)

12. Bacteria respond to O_2 in different ways. Indicate after the descriptions below <u>what type of bacteria</u> are being defined and give a <u>genus as an example</u>.

a) O_2 tolerant, respires and ferments

b) O_2 tolerant, ferments but sometimes carries out an anaerobic respiration

c) requires O_2, respires only

d) O_2 tolerant, ferments only

13. Bacteria have their maximum rate of growth at what temperature?

14. What is the optimum growth temperature?

15. What is the minimum growth temperature?

16. What is the maximum growth temperature?

17. What kind of bacteria are they?

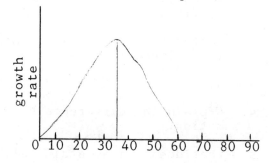

18. Indicate what kind of a fermentation is likely.

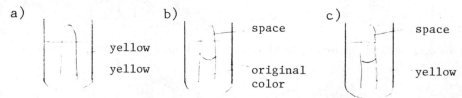

a) yellow yellow

b) space original color

c) space yellow

19. What are the waste products in an aerobic respiration?

20. What are the waste products in an anaerobic respiration where SO_4^{-2} is the electron and proton acceptor?

21. What is the maximum number of ATP that can be made in a respiration of a glucose molecule?

22. Name the compound that is metabolized and gives rise to the product below.

a) amines   b) indole
c) H_2S d) lactic acid

23. What is the name of the important pathway where citric acid is found?

24. What is the name of the pathway that converts glucose to 2 pyruvic acids?

25. In bacteria, where is the electron transport system found and what is its major function?

    a)   b)

26. Name the <u>genus</u> and <u>species</u> of bacteria that are responsible for the following diseases:

    a) Q-fever   b) mastitis
    c) dysentery   d) cholera
    e) brucellosis   f) typhoid fever
    g) gastroenteritis   h) tuberculosis in cattle

C.   BACTERIOLOGY SECOND LAB EXAM

1.   Indicate the types of molecules that starch, casein and gelatin are broken down to and the name of the enzyme.

	molecule	enzyme
casein		
starch		
gelatin		

2.   What is the term used to describe extracellular (outside the cell) enzymes?

3.   What is the term that means to break down a molecule with the addition of water?

4.   List 4 different types of fermentation

   a)   b)
   c)   d)

5.   List the 3 waste products generally produced during a fermentation.

   a)   b)   c)

6.   What is the net number of ATP made in the fermentation of a glucose molecule?

7.   If the methyl red test is positive (red), what kind of a fermentation pathway is likely?

8.   If the Voges-Proskauer test is positive (red), what kind of a fermentation pathway is likely?

9.   What kind of a molecule is NAD^+?

10. What does NAO^+ do?

11. Indicate what kind of a fermentation is likely.

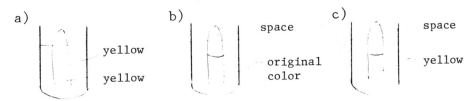

a) yellow / yellow   b) space / original color   c) space / yellow

245

12. What are the waste products in an aerobic respiration?

13. What are the waste products in an anaerobic respiration where SO_4^{-2} is the electron and proton acceptor?

14. What is the maximum number of ATP that can be made in a respiration of a glucose molecule?

15. Name the compound that is metabolized and gives rise to the product below.

    a) amines   b) indole
    c) H_2S d) lactic acid

16. What is the name of the important pathway where citric acid is found?

17. What is the name of the pathway that converts glucose to 2 pyruvic acids?

18. In bacteria, where is the electron transport system found and what is its major function?

    a)   b)

19. Name the genus and species of bacteria that are responsible for the following diseases:

    a) Q-fever   b) mastitis
    c) dysentery   d) cholera
    e) brucellosis   f) typhoid fever
    g) gastroenteritis   h) tuberculosis in cattle

20. Write the terms used to describe bacteria that have their maximum growth rate between the following temperatures.

    a) 0°C - 15°C
    b) 15°C - 45°C
    c) 45°C - 75°C

21. Write the terms used to describe the bacteria that respond to O_2 in the following ways. Also, give the genus of a bacterium.

    a) killed by oxygen, generally ferments but sometimes carry out anaerobic respiration

    b) ferments and respires aerobically

    c) oxygen tolerant but only ferments

    d) only respires aerobically

22. What can happen if you dilute bacteria in sterile pure water?

23. Why don't bacteria and other microorganisms grow well on dried foods or jams and jellies?

24. At what pH do most bacteria grow best?

25. Do bacteria or fungi grow best at low pH's?

C. BACTERIOLOGY THIRD LAB EXAM

TRUE-FALSE: Mark a T or F in front of each statement as is most approprite.

1. ____ Blackening of SIM medium cultures indicates tryptophanase activity.

2. ____ The VP test must not be performed before the MR test in the same culture tube.

3. ____ MRVP broth is essentially a glucose-peptone broth without added pH indicator.

4. ____ Koser's citrate broth is an example of a chemically defined medium.

5. ____ The development of a red color in a MRVP broth culture on addition of Kovac's reagent is interpreted as a positive VP test.

6. ____ The detection of an iron sulfide product in SIM medium implies a culture can perform anaerobic respiration.

7. ____ A sugar containing medium should not be used for the catalase test because of the possibility of acid production.

8. ____ Hektoen agar is differential for lactose fermenters and selective against gram-negative bacteria.

9. ____ EMB is selective against gram-positive bacteria and is differential for H_2S producers.

10. ____ A shallow layer of nitrogen-free mannitol mineral salts broth would be a suitable selective enrichment medium for the isolation of Rhizobium from soil.

11. ____ SIM stands for sulfate-indole-motility.

12. ____ The MR test detects the production of acetoin from glucose.

13. ____ Alpha-naphthol will detect acetylmethylcarbinol only under alkaline conditions.

14. ____ Nitrogen-fixation is an anaerobic process.

SHORT ANSWER QUESTIONS:

1. Briefly describe the basic steps in the "spore selection technique?"

2. Describe how blood agar can act as a differential plating medium?

3. Give at least 2 distinct reasons why a bacterial plate count rarely (if ever) gives an accurate estimate of the total number of viable microorganisms per milliliter (or gram) of a natural specimen.

CALCULATION PROBLEM

1. A bacterial plate count was performed on a specimen of spoiled meat according to the following protocol:

Step 1:   5 gm of meat was pulverized and diluted in a 95.0 ml dilution blank.

Step 2:   1.0 ml of suspension from step 1 was diluted further in a 49.0 ml dilution blank.

Step 3:   0.1 ml of suspension from step 2 was diluted further in a 9.9 ml dilution blank.

Step 4:   0.25 ml samples of suspension from step 3 were plated on each of 3 plates of PCA medium and the plates were incubated at 37°C for 48 hours.

Step 5:   Plates were counted and found to have 260, 370 and 280 colonies, respectively.

Now: Calculate the average viable cell count per ml of suspension in step 3?

Finally: Calculate the average viable cell count per gm of the original spoiled meat specimen.

C.   BACTERIOLOGY FOURTH LAB EXAM

TRUE-FALSE: Circle the T if the statement is completely accurate;
circle the F if the statement is inaccurate in any
significant part.

1.   T   F   Ammonification causes a net loss of nitrogen from the soil
as a volatile waste product.

2.   T   F   Organisms which perform denitrification in the soil are
able to generate useful chemical energy for growth by
means of anaerobic respiration.

3.   T   F   The genus of symbiotic nitrogen-fixing bacteria which
associate with leguminous plants is known as Bacteroids.

4.   T   F   In the root nodules of leguminous plants, nitrogen-fixing
bacteria produce a hemoglobin-like pigment.

5.   T   F   Azotobacter is an example of a strictly aerobic
nonsymbiotic nitrogen-fixing genus of bacteria.

6.   T   F   Fecal streptococci are sought in some countries and
states in water supplies as an indicator of fecal
pollution because of the frequency with which these
organisms cause human disease.

7.   T   F   Several tubes of lactose broth are inoculted in the
presumptive test for water pollution in order to detect
and estimte the "most probable number" of coliforms per
100 ml of water sample.

8.   T   F   The presumptive test for water pollution by coliforms is
considered positive if any one or more lactose broth
cultures show evidence of acid production.

9.   T   F   BGBL is considered an example of a selective culture
medium.

10.   T   F   The 3 most commonly sought indicator organisms of water
pollution are coliforms, fecal streptococci, and
Clostridium tetani.

MISCELLANEOUS

1.   Name 2 genera of disease-producing bacteria often transmitted
to man by polluted drinking water.

2.   What ingredient makes streptococcus agar a selective plating
medium?

3.   Lethal ultraviolet wavelengths center on what wavelength?

251

4. Which broad groupings (bacteria, blue-greens, fungi, algae, protozoa, plants, animals) possess the light repair mechanism for restoring cells damaged by UV-light exposure to an (essentially) normal state?

5. What enzymes are involved in the "dark repair" mechanism of restoring UV-damaged cells to an essentially normal state?

6. What determines the "host range" of a typical virus?

7. Viruses of bacteria are also (collectively) referred to as what?

8. There are 3 states in which a given virus can potentially exist. These would be:

   a)
   b)
   c)

9. A virus may be (rather simply) defined as: an _____ _____.

10. The numbers of infectious virus units (say, for a given bacterial virus) in a given suspension may be estimated by a process resembling a standard plate count for viable bacteria in a suspension. The calculations and most of the procedures are very analogous, except one counts infectious centers, or plaques, on a background lawn of bacterial cells.

    An adventuresome Bact. 221 student named Sal decided to perform a "plaque assay," as the procedure is called, in order to estimate the numbers of infectious virus units (= "plaque-forming units") per ml of a concentrated suspension of bacterial virus T4.

    She made a series of 5 ten-fold serial dilutions of her concentrated virus suspension. From the fifth serial dilution tube, she took 0.1 ml samples of diluted virus, mixed them with a young, susceptible host culture of E. coli in agar overlays and poured each virus-host overlay in a plate of TSA medium. After incubation of the 3 plates for 24 hours at 37°C, she counted the numbers of plaques appearing on each plate, with the following results: 205, 230, and 255 plaques, respectively.

    Use these data to calculate the number of plaque-forming units/ ml of the undiluted T4 virus suspension.

11. In addition to having lethal effects on cells, certain wavelengths of UV-light are also considered mutagenic. In which mechanism of repair of UV-induced damage does the possibility for mutations occur?

C. BACTERIOLOGY FINAL LAB EXAM

SHORT-ANSWER QUESTIONS: Keep answers <u>brief</u>.

1.   Based on the kinds of studies and the types of culture you have made of your class unknown, which of the following terms most certainly do <u>not</u> apply to your unknown (indicate after <u>each</u> entry, your responses; take into account the kinds of study/culture that we have not performed as well)?

    a. thermophile _____
    b. acidophile _____
    c. strict anaerobe _____
    d. halophile _____

2.   In performing the test for starch hydrolysis, a student tested his unknown by pouring Gram's iodine over an agar plate culture after allowing the organism to grow for a suitable period. Since the medium did not turn purple, the student concluded that the unknown hydrolyzed starch. His instructor told him he had made an error; explain the probable error.

3.   What is special about the digestive enzymes that some bacteria possess for the breakdown of large MW nutrients like starch, casein, and gelatin?

4.   In performing the MR and VP tests, a student first observed a red color in the VP test; then she added the MR test reagent and also observed a red color. Although she reported her unknown as being positive for both tests, her instructor suggested she try doing these tests over correctly. What was her error (explain why) and indicate how to correct it?

5.   Explain why it is important to check a culture, in a special test medium that is giving a negative reaction for a test activity, to make sure it is showing signs of growth.

6.   What <u>visual</u> observation must be made (<u>include mention</u> of specific test medium in use) before one concludes that an unknown bacterium:

a. hydrolyzes casein

b. produces indole

c. utilizes citrate as its sole organic carbon source

d. does <u>not</u> decarboxylate lysine

e. ferments glucose but does not produce gaseous wastes

f. produces hydrogen sulfide from methionine/cystine/cysteine digestion

253

g. does not produce detectable acetoin

h. is a strict (obligate) aerobe

7. Is it possible that an organism which actively ferments glucose would not ferment lactose? Explain your answer.

8. For each of the following activities of microbes, indicate whether a specific <u>indicator substance</u> is in the culture medium or must be added after incubation (<u>specify</u> the indicator/note "NONE" if there is no indicator substance):

a. indole production

b. citrate utilization

c. gelatin hydrolysis

d. hydrogen sulfide production

e. mannitol fermentation

f. growth as a facultative anaerobe

g. produces acetoin

9. Note a reasonable, specific location in nature where you would be likely to detect halophilic bacteria.

10. Write the chemical reaction catalyzed by the enzyme catalase.

Most enzyme activities serve a useful purpose for cells; what is the value to the cell of having catalase activity?

11. You have a TSA slant culture of your unknown; describe the oxidase test (first, what would you do; second, what observation would be interpreted as a "Positive" oxidase reaction?).

12. Differential and selective types of culture media are intended to make it easier to isolte specific types of bacteria from natural mixed specimens than if you used general purpose culture media (TSA).

Explain, in <u>general</u> <u>terms</u>, the difference between a medium that is primarily <u>differential</u> and a medium tht is primarily <u>selective</u> in function (i.e., how does each type of medium increase your ability to easily/rapidly isolate specific bacteria from natural specimens?).

Mannitol-salt agar is both differential and selective: explain <u>how</u> it can be both - be <u>specific</u>.

13. Sauerkraut is made by the fermention of _____.
    Several microbial species are known to be participants in this
    natural fermentation; where do they come from?

    What are 2 different purposes (practical) of natural
    fermentations, such as this one, which benefit man?

14. When a bacteriology student inoculates a peptone-causing broth
    and, following incubation, tests the culture for free ammonia,
    what reagent must be mixed with a sample of the culture and
    what color development occurs that is interpreted as positive
    evidence of free ammonia (what is importance of a sterile
    negative control in making this interpretation). Three parts
    to this answer.

15. In the tests for nitrate (NO_3-) reduction, what is the
    interpretation at each of the following stages (observations
    indicated)?

    a. Durham tube shows no bubble

    b. No gas evolved (no bubbles) when sulfamic acid added

    c. Gas evolved when powdered Zn added

16. What are bacteroids (kind of organism, where found, function)?
    microscopic appearance?

17. What chemical reaction is catalyzed by the enzyme urease?

    Many microbes in soil possess this type of enzyme; how does
    agriculture specifically use their (the microbes) ability to
    advantage?

18. The standard plate count procedure was used to estimate the
    number of aerobic, free-living, nitrogen-fixing bacteria
    (mesophilic) present in a particular soil sample. A series of
    10 progressive 1/10 (= 10-fold) dilutions of the soil were made
    in nitrogen-free buffer (sterile diluent) and 0.1 ml samples
    were plated in triplicate from each dilution tube with
    nitrogen-free culture medium (nitrogen-free means no organic
    nitrogen-containing compounds and no inorganic nitrogen-
    containing salts or ions are present - the only source of
    nitrogen is gaseous elementary nitrogen). The plates were
    incubated aerobically at room (ambient) temperature for 1 week
    and colony counts were performed. The following table
    summarizes some of the colony counts obtained:

Dilution tube	Colony counts obtained		
sampled	Plate 1	Plate 2	Plate 3
#4	too many to count on each of these plates		
#5	600	750	420
#6	80	68	92
#7	10	4	7
#8	0	0	1

Use the most appropriate set of data to calculate the average viable count of free-living, aerobic, nitrogen-fixing bacteria present in each gram of the soil sample. (Assume 1 gm of soil occupies 1 cc of volume.) Show your calculations/partial credit is possible.

TRUE-FALSE: Circle the T or F in front of each statement.

19. T F Coliforms are gram-negative, nonspore-forming bacteria, rod-shaped, which ferment sucrose with the production of acid and gas.

20. T F The detection of coliforms in a drinking water supply means the water contains infectious disease bacteria/viruses as well.

21. T F Enterobacter aerogenes is an example of a coliform.

22. T F Microbial disease agents commonly transmitted via fecally contaminated drinking water to humans include Mycobacterium tuberculosis.

23. T F If the presumptive test is positive and the confirmed test is negative, the lab will report the presence of coliforms in the water specimen that are inhibited by brilliant green/bile salts.

24. T F MPN stands for "most probable number" of disease bacteria present per 100 ml of a water specimen.

25. T F The membrane filter technique allows the direct estimation of the concentration of coliforms in a polluted water from a count of the "coliform" colonies that grow on the filter after incubation with Endo medium.

26. T F Another name for bacterial viruses is viroid.

27. T F A given virus has a specific of host species which is
           determined by the types of surface chemical receptors it
           has for attachement.

28. T F Once a virus particle is inside a host cell, it
           multiplied by a process of binary fission in that
           protected environment.

29. T F Viruses are not cellular forms; therefore, they cannot
           be described either a procaryotic or eucaryotic.

30. T F In the plaque assay of a particular virus preparation, 90
           plaques were detected when a $1/10^4$ dilution was mixed
           with host cells and plated. If 0.1 ml of the dilution
           was used on the plate, there must be approximately 9×10^5
           plaque forming units of virus in each ml of the undiluted
           preparation.

31. T F Mycelial fungi do not ordinarily consist of single cell
           populations.

32. T F Oral and vaginal forms of candidiasis (moniliasis) in
           humans are both caused by a yeast.

33. T F Beer and wine fermentations are carried out by a yeast of
           the genus <u>Saccharomyces</u>.

34. T F Yogurt, sauerkraut and kefir fermentations are carried
           out by a yeast of the genus <u>Saccharomyces</u>.

35. T F Agglutination reactions may be used in the laboratory for
           the rapid identification of some suspected bacterial
           disease agents as soon as they are isolated from the
           patient and cultured.

36. T F If antibodies for a specific known bacterium are detected
           in a serum specimen, it can be concluded that the person
           (or animal) has an active infection due to that
           bacterium.

37. T F Drug sensitivity tests, such as we tried to carry out,
           provide the physician attending a patient with a means of
           predicting which drugs are most likely to be useful in
           treatment to eradicate an infectious bacterium.

38. T F The presence of a zone of growth inhibition around a
           particular drug disk is interpreted as meaning that
           organism would be sensitive to that drug, when it is
           given to the infected patient.

39. T F True hemolyis involves a greenish discoloration of blood
           agar around colonies producing an enzyme known as a
           hemolysin.

40. T F True hemolysis is also known as "beta-hemolysis."

41. T F Bacteria which do not affect the red blood cells of blood agar cannot grow on this medium.

42. T F Any microorganism detected in a throat specimen from a clinically healthy individual is considered part of the normal microbiological flora of that person's throat.

43. T F The blood and heart are normally sterile.